AC/DC

Fig. 4.—Experiments in Killing Animals by the Alternating Current, as Conducted in the Edison Laboratory at Orange, N. J.

AC/DC

The Savage Tale of the First Standards War

Tom McNichol

Published by Jossey-Bass
A Wiley Imprint
989 Market Street, San Francisco, CA 94103-1741 www.josseybass.com

Library of Congress Cataloging-in-Publication Data

McNichol, Tom.
AC/DC : the savage tale of the first standards war / Tom McNichol.
p. cm.
Includes bibliographical references and index.
ISBN-13: 978-0-7879-8267-6
ISBN-10: 0-7879-8267-9
1. Electric currents, Alternating—History. 2. Electric currents, Direct—History.
3. Electricity—Standards—History. 4. Electricity—History. I. Title.
QC641.M36 2006
621.319'1309—dc22 2006013041

Printed in the United States of America
FIRST EDITION
HB Printing 10 9 8 7 6 5 4 3 2 1

Contents

AC/DC

Prologue

NEGATIVE AND POSITIVE

I've always had a healthy respect for electricity. Twice, it almost did me in.

The first time was serious. I was eleven years old, hanging out with my friend Mike in his basement. We had liberated some of his father's tools from a chest and were happily drilling, hammering, and sawing away the afternoon. I picked up a staple gun, which I had never used before, and began firing wildly like a Wild West gunslinger. There was a powerful recoil every time you shot a staple, so it seemed like you were doing something significant when you squeezed the trigger.

Looking around, I noticed that some insulation in the ceiling was sagging a bit—nothing a dozen well-placed staples couldn't fix. I dragged a metal chair under the spot, climbed on top, and with one arm stretched over my head Statue of Liberty style, began shooting staples into the insulation. It was difficult to aim while balancing on the chair, and one of the staples became embedded in a dark brown cord that ran along the edge of the ceiling. I'll just pull that staple out with my hand, I thought.

The brown cord turned out to be a wire buzzing with 120 volts of electricity, the standard household current in the United States. When I touched the metal staple rooted in the wire, my body became part of the electrical circuit. The current raced into my hand, down my arm, across my chest, down my legs, through the metal chair and into the ground—all at nearly the speed of light.

The sensation of having electricity course through your body is hard to put into words. Benjamin Franklin, who was once badly shocked by electricity (though not while flying a kite), described the feeling in a letter to a friend: "I then felt what I know not how to describe," Franklin wrote. "A universal blow throughout my whole body from head to foot, which seemed within as well as without."

A blow that seemed "within as well as without": yes. To me, the shock felt as though it was not simply running along the surface of my skin but was burrowing deep inside my body. The current felt like hot metal had been poured into my veins, a powerful surge that raced into the bones and down the marrow. The electricity was entering my body through my hand, but it didn't feel like the current had any particular location. It was everywhere. It was me.

The electricity flowing through my body was encountering resistance, which in turn was converted to heat. When people talk about criminals being "fried" in the electric chair, it's a fairly accurate description of what actually happens. I was slowly but steadily being cooked alive.

I'm not sure how long my hand clutched the electrified staple. Perhaps only a few seconds; maybe longer. Time seemed to have a different quality while in electricity's grip. The burst of current contracted the muscles in my hand, causing me to grasp the staple even harder, a phenomenon noted by Italian physician Luigi Galvani in the late eighteenth century when he touched an exposed nerve of a dead frog with an electrostatically charged scalpel and saw the frog's leg kick.

When a human touches a live wire, electricity often causes the muscles in the hand to contract involuntarily, an unlucky condition known among electrical workers as being "frozen on the circuit." Victims frozen on the circuit often have to be forcibly removed from the wire since they're unable to exercise control over their own muscles.

I was lucky. Just as my fingers were curling into a tight fist around the hot electrified staple, the sharp contraction of the muscles in my arm jerked my hand free. I immediately fell to the floor—pale, panting, and dazed, but otherwise uninjured. I had just felt the power of AC, or alternating current, the type of electricity found in every wall outlet in the home. In an AC circuit, the current alternates direction, flowing first one way and then the other, flipping back and forth through the wire dozens of times per second.

The 120 volts of electrical pressure that come out of an AC wall outlet are more than sufficient to kill a human being under the right circumstances. More than four hundred Americans are killed accidentally by electricity every year, and electric shock is the fifth leading cause of occupational death in the United States. And yet alternating current is utterly indispensable to modern life. The world as we know it simply couldn't do without AC power. Every light bulb, television, desktop computer, traffic signal, toaster, cash register, refrigerator, and ATM is powered by alternating current. The Information Age is built squarely on a foundation of electricity; without electric power, bits can't move, and information can't flow. Even the bits themselves are tiny electrical charges; a computer processes information by turning small packets of electricity on and off.

My second encounter with electricity's dark side wasn't quite as serious, but still left its mark. I was in college trying to jump-start my car on a frigid day, and had just attached the jumper cables to the battery of another car. As I moved to clamp the other end of the cables onto the dead battery, I stumbled and inadvertently brought the two metal clamps together. Once again, I had completed an electrical circuit, and once more, I was caught in the middle of it. A brilliant yellow-blue spark leaped from the cables, accompanied by a loud "pop." I immediately dropped the cables and discovered a black burn mark on my hand the size of a quarter, a battle scar from the electrical wars.

This time, I had been done in by DC, or direct current, the kind of current produced by batteries. Direct current moves in only one direction, from the positive to the negative terminal, but beside that, DC is the same "stuff" as AC: a flow of charged particles. A car battery produces about 12 volts of electrical pressure, only one-tenth the power that comes out of an AC wall outlet, but that didn't make my hand feel any better. Under the right conditions, direct current is every bit as deadly as alternating current.

And yet DC is also utterly essential to contemporary life. Every automobile on the road depends on DC to operate, along with every cell phone, laptop computer, camera, and portable music device. The same force that strikes people dead in lightning storms also saves lives. Cardiac defibrillators deliver a controlled burst of direct current to heart attack victims, forcing the heart muscles to contract and resume a regular rhythm.

Life and death, negative and positive. Electricity has many dualities, so it's only fitting that the struggle to electrify the world would give birth to twins: AC and DC. Long before there was VHS versus Betamax, Windows versus Macintosh, or Blu-ray versus HD DVD formats, the first and nastiest standards war of them all was fought between AC and DC. The late-nineteenth-century battle over whether alternating or direct current would be the standard for transmitting electricity around the world changed the lives of billions of people, shaped the modern technological age, and set the stage for all standards wars to follow. The wizards of the Digital Age have taken the lesson of the original AC/DC war to heart: control an invention's technical standard and you control the market.

The AC/DC showdown—which came to be known as "the war of the currents"—began as a rather straightforward conflict between technical standards, a battle of competing methods to deliver essentially the same product, electricity. But the skirmish soon metastasized into something bigger and darker.

In the AC/DC battle, the worst aspects of human nature somehow got caught up in the wires, a silent, deadly flow of arrogance, vanity, and cruelty. Following the path of least resistance, the war of the currents soon settled around that most primal of human emotions: fear. As a result, the AC/DC war serves as a cautionary tale for the Information Age, which produces ever more arcane disputes over technical standards. In a standards war, the appeal is always to fear, whether it's the fear of being killed, as it was in the AC/DC battle, or the palpable dread of the computer age, the fear of being left behind.

1

FIRST SPARKS

The story of electricity begins with a bang, the biggest of them all. The unimaginably enormous event that created the universe nearly 14 billion years ago gave birth to matter, energy, and time itself. The Big Bang was not an explosion in space but of space itself, a cataclysm occurring everywhere at once. In the milliseconds following the Big Bang, matter was formed from elementary particles, some of which carried a positive or negative charge. Electricity was born the moment these charged particles took form.

All matter in the universe contains electricity, the opposing charges that bind atoms together. Even human beings are awash in it; the central nervous system is a vast neuroelectrical network that transmits electrical impulses across nerve endings to the body's muscles and organs.

However, electricity, like the face of the Creator, is normally hidden from view. Most matter contains a balance of positive and negative charges, a stalemated tug-of-war that prevents electricity from manifesting itself. Only when these charges are out of balance do electrons move to restore the equilibrium, allowing electricity to show its face.

Electrical current is the flow of negatively charged electrons from one place to another in order to restore the natural balance of charge. It would take untold years and thousands of lives before humans learned to harness that flow and make those unseen charged particles do their bidding. Even then, electricity remained shrouded in mystery, an eccentric, invisible force with powers that seemed to come from another world.

Electricity first showed itself on earth as lightning, and as such, may have provided the original spark for life. Cosmologists believe that lightning may have provided some of the energy that transformed simple elements such as carbon, hydrogen, oxygen, and nitrogen into amino acids, the more complex molecular chains that are the building blocks of life.

Billions of years ago, the primordial surface of the earth was subjected to almost constant lightning strikes. Lightning is discharged when charged particles in the clouds separate; the lower portion of the cloud becomes negatively charged, producing an enormous electrical difference between it and the positively charged ground. The imbalance is discharged as a spark: lightning. A lightning bolt is a bundle of heat and energy, hotter than the surface of the sun and carrying an electrical force of more than a billion volts.

Lightning may have not only sparked organic life but also preserved plant life during crucial evolutionary choke points when fuel supplies ran low. During the Archaean age two billion years ago, carbon dioxide levels fell dramatically, drying up the supply of nitrates, which are essential for plant growth. Lightning is believed to have helped produce additional nitrates in the atmosphere, allowing plants to survive through this period. When plants began to flourish again, more oxygen was produced, making the earth increasingly suitable for animals, and later, humans. In many ways, we are the products of lightning, the sons and daughters of electricity.

The first humans knew nothing of lightning's creative power, only its terrible capacity for destruction. A jagged bolt from the heavens could incinerate someone in midstride, instantly turning a human being into a charred corpse. It was not the sort of power to be taken lightly. It would take millennia for humans to learn how to shield themselves from lightning, and longer still to learn its life-giving power. Lightning strikes sparked fires, which in time were controlled and put to use to cook food, provide warmth, and ward off dangerous animals.

The first creatures to put electricity to work were *Homo habilis*, or "Handy Man," the Stone Age humans that inhabited Africa about 1.8 million years ago. Handy Man, it turns out, wasn't all that handy. He hadn't yet worked out how to make fire; instead he waited for lightning to strike a bush or tree, and then carefully tended the flame. When it was time for the tribe to move to another location, Handy Man took lit branches along to start a new fire, or simply waited for lightning to strike again somewhere else.

For *Homo sapiens*, lightning and electricity would likewise be a luminous mystery. Around 600 B.C., the Greeks discovered that amber, a soft golden gem formed from fossilized tree sap, behaved oddly when rubbed by a piece of fur: the stone attracted pieces of straw or hair. Sometimes, the amber would even emit a spark, a miniature lightning bolt. The science behind this strange effect would remain a mystery for more than two thousand years, but the Greeks had discovered static electricity. As we now know, the fur transferred negatively charged electrons to the amber, giving it an imbalanced charge, which in turn attracted the straw. The phenomenon would later give electricity its name: *elecktron* is the Greek word for amber.

Even as humans struggled to understand electricity, the subject continued to be clouded by superstition. Thales of Miletus, an early Greek philosopher and mathematician, interpreted the curious properties of amber as evidence that objects were alive and possessed immortal souls. Greek mythology explained electricity by associating lightning with Zeus, the supreme god, who threw bolts of lightning down from the heavens to vent his anger at enemies below. Virgil's *Aeneid* recounts the tale of Ajax, who, boasting of his own power, defied lightning to strike him down. Such a dare amounted to nothing less than shaking his fist in the face of the gods, and led to a predictably unhappy ending. In short order, Ajax was felled by an expertly aimed lightning bolt from the sky.

Lightning was so fearsome that many cultures sought to ascribe meaning to what seemed like a wantonly destructive power. The Etruscans and Romans believed that lightning was not simply a weapon of the gods but a message from them. The Etruscans were particularly keen observers of lightning, dividing the sky into sixteen sections in order to determine the significance of a bolt. Lightning moving from west to north was considered disastrous, while lightning to the left hand of the observer was thought to be fortunate. The Etruscans even compiled a sacred book about the art of interpreting lightning strikes, and laid out their towns in accordance with signs gleaned from the heavens.

In Roman times, objects or places struck by lightning were considered holy. Roman temples often were erected at these sites, where the gods were worshipped in an attempt to appease them. A man struck by lightning who lived to tell the tale was considered someone especially favored by the gods. In most cases, however, lightning was utterly destructive. A thunderbolt, the Roman poet Lucretius wrote, "can split towers asunder, overturn houses, tear out beams and rafters, move monuments of men, struck down and shattered, rob human beings of life, and slaughter cattle."

Lightning mythology readily spread to other cultures—the phenomenon was clearly something that demanded explanation. The Vikings believed lightning was caused by Thor striking a hammer on an anvil as he rode his chariot across the sky. In Africa, Bantu tribesmen worshipped the bird-god Umpundulo, who directed lightning. Medicine men were sent into storms to bid Umpundulo to strike far away from a village, a practice that continues to this day in parts of Africa. The Book of Job places lightning in the hands of a wrathful God: "He fills his hands with lightning and commands it to strike its mark." The Koran states that lightning, which is directed by Allah, can be a force for both creation and destruction: "He it is who shows you the lightning causing fear and hope."

Native American tribes were particularly attuned to lightning's dual nature, its power to kill and to give birth. Native tribes saw with remarkable clarity the inherent duality of electricity centuries before Western science would describe electrical current as a flow between negative and positive poles. One legend has Black Elk, an Oglala Sioux, testifying: "When a vision comes from the thunder beings of the West, it comes with terror like a thunder storm; but when the storm of vision has passed, the world is greener and happier; for wherever the truth of vision comes upon the world, it is like a rain. The world, you see, is happier after the terror of the storm. . . . You have noticed that truth comes into this world with two faces. One is sad with suffering, and the other laughs; but it is the same face, laughing or weeping."

Negative and positive, plus and minus, good and evil, life and death. The Chinese Taoists termed the pair of opposites found in nature yin and yang, and the concept is well suited to electricity. Yin and yang are not opposites in conflict; they are simply different aspects of the same system. One depends on the other for its existence. As one aspect overcomes the other, the seeds of a reversal are sown.

Likewise, the negative and positive poles in electricity represent an ever-changing polarity—the dominance of a negative charge contains the inception of a rise of a positive charge. The famous yin-yang symbol expresses the concept with elegant simplicity: the blackest part of the symbol contains a tiny white dot, and the whitest part a black dot, the seeds of the inevitable opposite about to give birth.

Not until the end of the Middle Ages would philosophers begin to look at electricity scientifically. The first truly scientific study of electricity and magnetism was taken up by William Gilbert, an English physician to Queen Elizabeth I. Gilbert's book *De Magnete* (On the Magnet), published in Latin in 1600, introduced the term *electricity* to describe the attractive force of rubbed amber.

Gilbert spent seventeen years experimenting with magnetism and electricity, attempting to strip away the myths that had shadowed electricity since the dawn of time. Gilbert was the first to describe a relationship between electricity and magnetism, as well as being the originator of the terms *electric force*, *magnetic pole*, and *electric attraction*. Gilbert divided objects into "electrics" (such as amber) and "non-electrics" (such as glass). He attributed the electrification of an object to the removal of a fluid, or "humour," which then left an "effluvium," or atmosphere, around the body. Gilbert actually wasn't far off the mark. His "electrics" would later be known as *conductors*, while the "non-electrics" would be called *insulators*. The "humour" that was stripped off objects would be known as a "charge" and the "effluvium" that was created became an "electric field."

Before long, experimenters developed machines that could produce large amounts of static electricity on demand. In 1660, German experimenter Otto von Guericke made the first electrostatic generator out of a ball of sulfur and some cloth. The sulfur ball was mounted on a shaft placed inside a glass globe. A crank rotated the ball against the cloth, and a static electric spark was produced. To von Guericke, the sulfur ball symbolized the earth, which shed part of its electric "soul" when rubbed—not exactly a scientific explanation. But the machine worked, letting experimenters produce electric sparks whenever they wanted.

In 1745, Pieter van Musschenbroek, a physicist and mathematician in Leiden, Holland, was one of several experimenters to fashion a device that would become known as the Leyden jar. Van Musschenbroek's Leyden jar consisted of a glass vial partially filled with water. A beaded metal chain dangled in the water, held by a wire that ran through a cork stopper and out the top of the jar, terminating in a metal knob. Van Musschenbroek held the jar in one hand and touched the knob to a spark generator. When nothing happened, van Musschenbroek touched the knob with his other hand, and at that instant, got the shock of his life:

"My right hand was struck with such force that my whole body quivered just like someone hit by lightning," van Musschenbroek wrote. "Generally the blow does not break the glass, no matter how thin it is, nor does it knock the hand away, but the arm and the entire body are affected so terribly I can't describe it. I thought I was done for."

Van Musschenbroek couldn't figure out what had caused the shock—after all, the jar was no longer connected to the static generator when he got zapped. He later told an associate he would never try such an experiment again, but others weren't so cautious. Leyden jar experimenters soon reported everything from nosebleeds, convulsions, and prolonged dizziness to temporary paralysis when they unleashed the charge with their hand.

The Leyden jar was electricity in a bottle, an ingenious way to store a static electric charge and release it at will. When a charge was applied to the inside surface of the Leyden jar, it meant that the outside surface (which was insulated from the inside) had an equal but opposite charge. When the inside and outside surfaces were connected by a conductor—in this case, a human hand—the circuit was completed, and the charge was released with a dramatic spark. The Leyden jar was the forerunner of what today is known as a capacitor. Capacitors are found in a camera's electronic flash, for example, used to store a charge and then release it instantly when a picture is snapped.

Eventually, the Leyden jar was refined so that the electric charge could be released without having to shock the user, a boon for further experimentation. Leyden jars quickly became as much a novelty item as a scientific instrument. Scores of enterprising experimenters drew rapt crowds all over Europe demonstrating electricity with the jars. They killed birds and small animals with a burst of stored electric charge and sent electrostatic sparks through long wires over rivers and lakes. In 1746, Jean-Antoine Nollet, a French clergyman and physicist, discharged a Leyden jar in the presence of King Louis XV, sending a current of static electricity rushing through a chain of 180 Royal Guards who were holding hands. In another demonstration, Nollet connected a row of Carthusian monks with a metal wire. A Leyden jar was used to send a charge through the wire, and the white-robed monks were said to have leapt simultaneously into the air, goosed by a jolt of electricity.

One of the electric showmen of the day was Dr. Archibald Spencer, a physician from Scotland who came to Boston in 1743 to demonstrate "electric magic" to an audience. Spencer's demonstrations were high on theatrics—in one display, he drew sparks from the feet of a boy hanging from the ceiling by silk cords. The audience was astonished, never having seen such wonders performed. One audience member was particularly fascinated by the demonstration, a visiting postmaster from Philadelphia named Ben Franklin.

2

LIGHTNING IN A BOTTLE

Ben Franklin flying a kite in a thunderstorm: It's an image burned in the brain of every American schoolchild, an icon as durable as Paul Revere galloping through the countryside or George Washington blithely chopping down a cherry tree. There stands Ben, usually in full colonial dress, tugging on the string of a kite that's being struck by a jagged bolt of lightning. A key tied to the end of the string gives off a faint but unmistakable glow. Franklin's face is curiously impassive, particularly for a man who's come within inches of several million volts of electricity.

Like many of history's most familiar scenes, the Franklin kite story is a blend of fact and fiction, what a modern-day movie advertisement might describe as being based on a true story. Franklin did indeed fly a kite in a thunderstorm to see whether lightning was a form of electricity, but he wasn't the first to test this theory, nor was his experiment a very smart approach—Franklin came perilously close to being incinerated on the spot. As it turns out, the kite demonstration was only the most celebrated of Franklin's many experiments with electricity during his lifetime. Had Ben never flown the kite, his contribution to the electrical arts would have been no less important.

Unlike almost every experimenter who would follow him, Franklin was only a part-time player in the field of electricity. Nearly all of Franklin's discoveries in electricity took place within a six-year period culminating with his kite experiment sometime in June 1752. Such was the expansiveness of Franklin's genius that he managed to squeeze groundbreaking electricity research into such a brief period, leaving time for him to be a publisher, writer, postmaster, statesman, raconteur, political philosopher, insurrectionist, and inventor (of the Franklin stove, bifocals, the flexible medical catheter, and swim fins).

Franklin caught the electricity bug after attending Archibald Spencer's demonstrations in Boston, a show that included drawing long sparks from a Leyden jar as well as from statically charged volunteers. "Being on a subject quite new to me, they equally surprised and pleased me," Franklin later wrote of Spencer's stunts. Franklin's only complaint was that Spencer wasn't much of a showman; the doctor's electrical tricks "were imperfectly performed, as he was not very expert."

Franklin's insatiable curiosity and theatrical flair made him a natural to take on the mysteries of electricity. Franklin also happened to have free time on his hands. He was in the process of selling his printing shop in Philadelphia and retiring from business in order to devote his time to what Franklin called "philosophical studies and amusements." After seeing Spencer's show, Franklin went out and purchased all the electrical equipment he could find, including a Leyden jar. Franklin also obtained a long glass tube for generating static charges, a gift from Peter Collinson, a botanist and fellow of the Royal Society of London. Collinson would quickly become Franklin's most trusted correspondent in matters relating to electricity, a sounding board for emerging theories. The two men exchanged dozens of letters, and Franklin's folksy, clear-headed descriptions of his experiments, which were later published, would demystify electricity for thousands.

Once Franklin committed himself to learning everything he could about electricity, he could barely contain his excitement. "For my own part, I never was before engaged in any study that so engrossed my attention and my time as this has lately done," Franklin wrote to Collinson. The fanciful tricks demonstrated by Dr. Spencer had appealed to Franklin's roguish nature, and he was soon entertaining friends with his own electrical stunts. Franklin applied an electrical charge to an iron fence surrounding his Philadelphia house so that the fence gave off a harmless but dramatic spark when it was touched. He fashioned a fake spider out of metal and then put a charge to it, making it scurry across the ground. He rigged a portrait of King George II so that anyone touching the king's crown received "a high-treason shock." He charged drinking glasses filled with wine so that unsuspecting guests were treated to a spark as they imbibed. He also participated in a parlor game called "the electric kiss," in which participants passed a charge around a circle with their lips.

In the summer of 1749, Franklin threw a "party of pleasure" on the banks of the Schuylkill River for his friends, with electricity as the featured attraction. Franklin described the affair: "A turkey is to be killed for our dinners by the electrical shock, and roasted by the electrical jack, before a fire kindled by the electrical bottle; while the healths of all the famous electricians in England, Holland, France, and Germany are to be drank in electrified bumpers, under the discharge of guns from the electrical battery." The electrified turkey, to the surprise of guests, proved to be quite tasty. "The birds killed in this manner eat uncommonly tender," Franklin wrote.

Franklin delighted in such antics, presenting his latest electrical trick to friends with a mischievous twinkle in his eye. Still, Franklin took the subject of electricity seriously. In his studies, he was guided by one of his favorite aphorisms: "The noblest question in the world is: 'What good can I do in it?'" Franklin wasn't so much interested in acquiring knowledge about electricity for its own sake; the goal was always to use the information for the good of all.

Franklin sought to understand electricity through rigorous experimentation, a somewhat novel approach at the time. He performed dozens of experiments with electrical charges drawn from a Leyden jar ("that wonderful bottle," Franklin called it) and soon began compiling a list of the peculiar properties of electricity. "Electric fire loves water, is strongly attracted by it," noted Franklin after seeing how water, and even dampness, was a particularly good conductor of electricity. Franklin also discovered—the hard way—that electricity doesn't merely travel along the surface of an object, but rather passes entirely through it. "If anyone should doubt whether the electric matter passes through the substance of bodies, or only over and along their surfaces, a shock from an electrified large glass jar, taken through his body, will probably convince him," Franklin wrote.

Experimenting with electricity was dangerous work, and Franklin received his share of unexpected shocks. One jolt was particularly harrowing. A few days before Christmas 1750, Franklin strung together two large Leyden jars, intending to kill a turkey with electricity for his holiday feast. Franklin inadvertently grasped the charged metal chain of one of the Leyden jars, thus completing the circuit. There was a brilliant flash of light and "a crack as loud as a pistol" as the jar discharged, sending a large burst of electrical charge through Franklin's body.

"The first thing I took notice of was a violent, quick shaking of my body, which gradually remitting, my sense was gradually returned," Franklin wrote. "That part of my hand and fingers which held the chain was left white, as though the blood had been driven out, and remained so eight or ten minutes after, feeling like dead flesh; and I had a numbness in my arms and the back of my neck, which continued to the next morning."

Despite producing some painful lessons, Franklin's experiments began to bear fruit. At the time, it was widely believed that electricity involved two kinds of fluids, known as vitreous and resinous, which operated independently of one another. These two types of fluid were meant to explain why some electrified objects attracted other substances, while others repelled them. Franklin's own experiments convinced him that electricity was instead a single fluid that manifested itself as two different charged states. As Franklin explained in a letter to Collinson, "Hence have arisen some new terms among us: we say B (and bodies like circumstanced) is electrized *positively;* A *negatively*. Or rather, B is electrized *plus;* A *minus*." Franklin apologized for the new terminology, adding, "These terms we may use until your philosophers give us better."

As it turned out, Franklin's terms—negative and positive—would stand the test of time, and persist to this day. Franklin's only mistake was stating that electricity flowed from positive—the terminal with an "excess" of charge—to negative, the terminal with a "shortage" of charge, when in fact it's the other way around. It would be nearly 150 years until the electron was discovered, the negatively charged particle whose movement is the basis of current flow. By that time, Franklin's original sense of positive and negative had been in use for so long that his terminology was retained. Even today, electrical circuits are drawn showing the electricity flowing from positive to negative, even though the electron flow is actually in the opposite direction.

Franklin may have gotten the direction of the flow wrong, but he was correct in viewing electricity as a flow of charge that moves in an effort to reach a state of equilibrium. As Franklin noted, when the top of a Leyden jar was charged positively, the bottom was charged negatively in exact proportion. Franklin's discovery of this phenomenon, known as conservation of charge, was an important breakthrough. Electricity, far from being some magical, capricious force, acted with the predictability of an accountant, always seeking to balance nature's ledger book of charge.

As Franklin began to piece together the laws that governed electricity, he never lost sight of searching for practical applications of his knowledge. Franklin found one area of inquiry particularly promising: "the wonderful effect of pointed bodies, both *drawing off* and *throwing off* the electrical fire," he wrote in another letter to Collinson.

"Points have a property, by which they draw on, as well as throw off the electrical fluid, at greater distances than blunt objects can," Franklin wrote. "Thus a pin held by the head, and the point presented to an electrified body, will draw off its atmosphere at a foot distance; where, if the head were presented instead of the point, no such effect would follow."

Franklin didn't understand exactly why pointed objects drew sparks better than blunt ones and, ever the pragmatist, didn't really care. "To know this power of points may possibly be of some use to mankind, though we should never be able to explain it," Franklin wrote. The power of pointed objects excited Franklin because he saw a useful way to exploit the phenomenon: as a way to draw lightning away from buildings. He noted that lightning, like the electricity in his experiments, seemed to be attracted to tall pointed objects: tall trees, the masts of ships, the spires of churches, and chimneys. Taking note of a sea captain's account of lightning striking his ship's mast, Franklin found it significant that the mast gave off sparks shortly before the bolt of lightning actually struck. The metal mast was drawing off a charge from the cloud just as Franklin had drawn off sparks with the pointed end of a pin in his laboratory.

Perhaps lightning was nothing more than a gigantic spark; an oversized version of the small sparks Franklin had discharged in his experiments hundreds of times. Franklin compiled a list of a dozen properties shared by electricity and lightning, including the color of the light emitted; its swift, crooked motion; its ability to be conducted by metals; its crack or noise in exploding; and its sulfurous smell. "Electrical fluid is attracted by points," Franklin wrote. "We do not know whether this property is in lightning. But since they all agree in particulars wherein we can already compare them, is it not probable they agree likewise in this?" To this question Franklin appended a brief declaration that would become a kind of battle cry for researchers to follow: "Let the experiment be made!"

To determine whether clouds that contain lightning are electrified or not, Franklin proposed a novel experiment: "On the top of some high tower or steeple, place a kind of sentry-box, big enough to contain a man and an electrical stand. From the middle of the stand let an iron rod rise and pass bending out of the door, and then upright twenty or thirty feet, pointed very sharp at the end. If the electrical stand be kept clean and dry, a man standing on it when such clouds are passing low might be electrified and afford sparks, the rod drawing fire to him from a cloud. If any danger to the man should be apprehended (though I think there would be none), let him stand on the floor of his box, and now and then bring near to the rod the loop of a wire that has one end fastened to the leads, he holding it by a wax handle; so the sparks if the rod is electrified, will strike from the rod to the wire and not affect him."

Franklin wasn't the first person to suggest that lightning was a form of electricity; he was, however, the first to propose a scientific method of proving the theory. Franklin never performed the experiment exactly as he proposed it, but the suggestion and the theory behind it attracted worldwide attention after his letters to Collinson were included in a 1751 pamphlet, *Experiments and Observations on Electricity*, which soon was translated into French, German, and Italian. The booklet caused a sensation in Europe, turning Franklin into an international celebrity, and sparking a surge of amateur experimenting with electricity. The most important of these put Franklin's proposed lightning experiment to the test.

On May 10, 1752, in the village of Marly-la-Ville just north of Paris, French experimenters constructed a sentry box according to Franklin's specifications, topped by a pointed iron bar, forty feet high. At twenty minutes past two in the afternoon, a storm cloud passed over the sentry box, and suddenly, the iron bar began attracting sparks of fire. No lightning had struck the iron bar; the metal was drawing off a charge from the storm cloud, just as Franklin had predicted. The experiment was soon replicated in several other locations throughout Europe, though not always with happy results. Georg Wilhelm Richmann, a Swedish physicist working in Russia, was killed by lightning while attempting to replicate the Franklin experiment. Richmann was found dead on the ground with a red spot on his forehead and two holes in his shoes, the entry and exit points of the electrical flow.

News traveled slowly in eighteenth-century America, and Franklin was unaware that the French already had performed his lightning experiment when, about a month later, he decided to try it himself. Thus, while the conventional tale has Franklin's experiment "proving" his theories about lightning, Franklin's concepts actually were confirmed experimentally a month before he picked up his kite.

The only detailed account of the kite experiment was written not by Franklin but by his friend Joseph Priestley, a renowned chemist who wrote about it fifteen years later. According to Priestley's account,

Franklin intended to perform the experiment in a sentry box constructed atop the steeple of Christ Church in Philadelphia. But the steeple's construction was delayed, and Franklin came up with a characteristically whimsical way to capture lightning: with a kite. Sometime in June 1752, Franklin fashioned a kite out of a large silk handkerchief stretched over two cross-sticks and fastened to a long length of hemp twine. To the top of the upright stick he attached a foot-long sharp-pointed wire; to the near end of the twine he tied a key, and knotted a silk ribbon below the key. Grasping the dry silk ribbon, Franklin stood under the awning of a small shed in the middle of a field, waiting for an approaching summer thunderstorm.

Franklin was not alone; accompanying him on the mission was his twenty-one-year-old son, William. (Most popular depictions of the kite experiment discreetly airbrush William out of the picture, and the few paintings that include him often get it wrong. A widely circulated Currier and Ives painting of the kite experiment, for example, shows William as a young boy. Franklin himself was forty-six when he flew the kite, but many drawings show him as an elderly, white-haired sage.)

According to Priestley's account, Franklin dreaded the ridicule of performing an unsuccessful experiment in public, so he kept the kite test to himself, making sure that his son was the only witness to the events of that June day. Some have seized on this secrecy as evidence that the kite experiment never actually happened, but there's little to support the notion that the experiment was faked. Franklin would later have a nasty falling out with his son William, the only witness to the experiment, but the young Franklin never disputed the official version of events.

Once the kite was aloft, there was a tantalizing pause before the heavy storm clouds moved in, a pre-electric tingle of anticipation. Priestley picks up the tale: "One very promising cloud had passed over it without any effect, when, at length, just as he was beginning to despair of his contrivance, he observed some loose threads of the hempen string to stand erect and to avoid one another, just as if they had been suspended on a common conductor. Struck with this promising appearance, he immediately presented his knuckle to the key, and (let the reader judge of the exquisite pleasure he must have felt at that very moment) the discovery was complete. He perceived a very evident electric spark. Others succeeded, even before the string was wet, so as to put the matter past all dispute, and when the rain had wet the string he collected electric fire very copiously."

Franklin ended the experiment there, satisfied that he had proved his point. Despite countless depictions to the contrary, no lightning bolt actually struck the kite directly. Had it done so, the experiment probably would have been Franklin's last. The charge Franklin had drawn from the sky had precisely the same properties as the electrostatic charges he had produced countless times in his laboratory, proof that lightning was indeed a form of electricity.

Franklin characteristically viewed the kite test as a means to an end. If lightning was electricity, then it could be "drawn off" by pointed metal just as Franklin had drawn off static charges in his laboratory. This led to one of Franklin's most valuable inventions: the lightning rod. The rod was a pointed piece of metal affixed to the highest point of a building; a metal wire attached to the rod ran down the side of the building and into the ground. When lightning struck the rod, the electricity ran down the wire and into the earth, preventing damage to the building.

Franklin's rods soon sprang up on roofs throughout the colonies and Europe, and *Poor Richard's Almanack* published instructions on how people could fashion their own lightning rods. The device would save countless lives and buildings, and Franklin himself considered it to be his most important invention. Franklin never patented his lightning rod, even though it would have made him a wealthy man. Seeing his scientific theories put to practical use was reward enough.

Franklin performed almost no electricity experiments after 1752, as his time was increasingly taken up by politics and the gathering storm of the American Revolution. He had packed a lifetime of electrical experimentation into a handful of years, and many of the electrical terms he coined would still be around centuries later:

positive and *negative charge*, *neutral*, *conductor*, and *condenser*. As one observer noted, when Franklin began his experiments, electricity was little more than a curiosity; he left it a science.

Still, it would be a science for tinkerers. Electricity had many deep mysteries, and often the experimenter scarcely knew where to begin digging. Franklin stumbled into many of his findings, guided by an intuitive sense of where to look. There were countless dead ends, but to the experimenter, failure is a small victory in itself, bringing the answer one step closer.

The next electrical giant to follow Franklin would be a master tinkerer: Michael Faraday. Born near London in 1791, Faraday was the son of a poor blacksmith. He was forced to drop out of school at age thirteen to help support his family, and he became an apprentice bookbinder. With access to thousands of books, Faraday taught himself everything there was to know about electricity. "Facts were important to me and saved me," Faraday would later note. "I could trust a fact, and always cross-examined an assertion." Faraday would make up for his lack of formal education with brilliantly conceived experiments that made him the foremost electricity researcher of his day, the king of the tinkerers.

Faraday lucked into a position at London's Royal Institution working as an assistant for Sir Humphrey Davy, and got to meet some of the leading electrical researchers of the day. He traveled to Milan to meet Alessandro Volta, who in 1799 created the first battery by stacking alternating copper and zinc rings and submerging them in an acid solution. The so-called voltaic pile produced electricity without needing to be charged like a Leyden jar—the direct current was generated by a chemical reaction between the metals and the acid.

Inspired by Volta and a wave of electrical experimenting in Europe, Faraday soon took up his own work. Many of Faraday's experiments probed the curious relationship between electricity and magnetism. Faraday discovered that when he moved a loop of a wire through a magnetic field, a small burst of current flowed through the loop momentarily, a phenomenon known as induction. Faraday's induction ring was the first electric transformer. In another series of experiments, Faraday discovered a way to produce a steady flow of current by attaching two wires to a copper disk and then rotating the disk between the poles of a horseshoe magnet. This was the first generator (or dynamo, as it was known in the nineteenth century)—a tiny factory of direct current. By reversing the principle of the generator, Faraday constructed the first electric motor. By 1831, the key elements of the coming Age of Electricity—the electric motor, generator, and transformer—had been established in Faraday's laboratory.

Ben Franklin always suspected that electricity had hidden uses, writing in 1750, "The beneficial uses of this electric fluid in the creation we are not yet well acquainted with, though doubtless such there are, and those very considerable." It had taken humans millions of years to view electricity as something to be studied rather than feared. Now, electricity was something not simply to study but perhaps to control.

But who had the power to control a flow of invisible particles? Electricity awaited its master.

3

ENTER THE WIZARD

On February 11, 1847, in the tiny village of Milan, Ohio, Nancy Edison gave birth to her seventh child, Thomas Alva Edison. The boy was born into a world lit by candles and gas lamps; before his life was over, the entire planet would be lit by the steady glow of electric lights that he invented.

Nancy Edison was already middle-aged when she gave birth to Thomas Alva, who would be known as "Al" through adolescence. At birth, young Al's head was so unusually large that the village doctor feared the child might have brain fever. Edison's enormous head turned out to be more metaphor than malady, a winking signal to the world of the oversized brain it contained.

The Edisons were solidly working class; Edison's father, Samuel, ran his own shingle factory and lumberyard. The clan's most prominent trait was a long-standing history of almost stupefying stubbornness. The Edisons (originally, the family pronounced it with a long "e": EE-di-son) were of Dutch and English stock who came to the New World and settled in New Jersey in the 1730s. When the Revolutionary War broke out, the Edisons obstinately supported the British Crown against the colonists, and the entire family was banished to Canada. In the 1830s, the Edisons once again backed the wrong horse, pushing for the overthrow of the Royal Canadian government. The family was again sent packing, this time to Ohio. As an adult, Thomas Edison eventually would settle in northern New Jersey, not far from where his ancestors could have stayed in the first place if they weren't quite so pig-headed. Edison inherited the primary family trait in spades; his stubbornness would be responsible for many of his greatest triumphs and several of his biggest mistakes.

One of Edison's first memories was of the drowning of a village boy in his hometown of Milan. At the age of five, Edison accompanied the boy to a gully on the outskirts of town in order to swim in a small creek. As Edison later recounted in notes to his official biographers, "After playing in the water awhile, the boy with me disappeared in the creek. I waited around for him to come up but as it was getting dark, I concluded to wait no longer and went home. Some time in the night I was awakened and asked about the boy. It seems the whole town was out with lanterns and had heard that I was last seen with him. I told them how I had waited and waited, etc. They went to the creek and pulled out his body."

Edison's coldly dispassionate description of the incident was later sanitized in his 1910 authorized biography, which recounts young Tom walking home from the drowning "puzzled and lonely, but silent as to the occurrence" with "a painful sense of being in some way implicated." In fact, Edison's indifference to pain—whether his own or a fellow creature's—would be a lifelong characteristic. While Edison could bestow sudden acts of kindness on hard-working employees or show childish enthusiasm for a new invention, misery of any kind left him cold. He had little interest in creature comforts or in the comfort of his fellow creatures. His view of suffering was oddly detached, as though he were observing a laboratory experiment.

Edison didn't enter school until age eight due to a bout with scarlet fever, and his scholastic career would prove to be decidedly brief. Edison spent about three months of his life in a classroom, and by all accounts, he hated every minute of it. The rote memorization of facts and dull drilling offered little to a boy whose natural curiosity about the world around him was unusually acute. Edison's teacher, noting the boy's utter indifference to his studies, thought he was "addled." Edison's father, Samuel, seemed to agree; he pulled his son out of the local elementary school, never to return.

Edison's mother, however, far from believing her son to be dull-witted, saw him as possessing an unusually sharp mind. Nancy Edison was a former schoolteacher and took up the task of educating her son at home, instilling in him a love of books and of knowledge for its own sake. "My mother was the making of me," Edison would later say. "She understood me, she let me follow my bent."

Early on, Edison was drawn to science. The first book he remembered reading all the way through was *School of Natural Philosophy*, which described simple chemistry experiments that could be done at home. Edison set up a makeshift chemistry lab in the cellar of his parents' home, lining the walls with more than two hundred small bottles of chemicals, each labeled "POISON" to ward away snooping adults. His parents would later recall hearing occasional muffled explosions from the cellar, a sign that young Al had learned another hard lesson about the combustibility of certain chemical combinations. Edison's preferred method of experimentation as a child was trial and error, and it would remain a characteristic strategy for the rest of his life. For Edison, a wrong answer simply meant that he was just a little closer to the correct one.

Besides chemistry, the young Edison was fascinated with electricity, particularly the electric telegraph. Samuel Morse had demonstrated the first practical telegraph nine years before Edison was born. Morse's device used pulses of direct current to deflect an electromagnet, which moved a marker to produce written codes on a strip of paper. By the time Edison was a boy, telegraph lines had begun to connect cities and towns across the country. Edison cobbled together a crude working telegraph out of scrap metal, powering the device with a voltaic cell battery. Being able to send invisible pulses of electricity down a wire and out the other end filled Edison with a profound sense of wonder. What was this strange thing called electricity? How did it really work? He posed the questions to anyone who would listen until one day he got back a satisfactory, if somewhat oblique answer. A traveler from Scotland told Edison that electricity was "like a long dog with its tail in Scotland and its head in London. When you pulled its tail in Edinburgh, it barked in London."

"I could understand that," Edison would recall, "but I never could get it through me what went through the dog or over the wire." The telegraph operators who sent Morse code messages hundreds of miles didn't really understand how electricity worked, either. Half a century later, Lord Kelvin, one of the leading physicists of his day, admitted that after a lifetime of studying electricity, he knew as little as when he had begun. The electron, the fundamental unit of electricity, wouldn't be discovered until 1897.

Edison took his first job at age twelve, selling newspapers and snacks aboard the bustling commuter trains of the Grand Trunk Railway that ran between Port Huron, Michigan (where the Edisons had moved) and Detroit. Edison quickly found a way to fulfill his duties as a "train boy" and keep up with his science experiments. He set up a laboratory in the mail car of the train, stocking his rolling lab with chemicals, test tubes, bales of wire, and voltaic batteries neatly arranged on shelves. For Edison, the job on the train was a perfect set-up; the work was easy and gave him plenty of spare time to follow his own curiosity. He gave his mother a dollar a week from his earnings, and then spent whatever was left on books and laboratory supplies.

"The happiest time of my life was when I was twelve years old," Edison later wrote. "I was just old enough to have a good time in the world, but not old enough to understand any of its troubles."

Just before Edison turned thirteen, he suffered an injury that would profoundly affect not only his life but also the course of science. There are differing accounts of the incident. In one version, Edison was about to be struck by a train while standing on a platform when he was lifted by his ears to safety. Another version has an angry train conductor ferociously boxing Edison's ears for causing a fire on the train with his chemicals. Whatever the cause of the injury, Edison remembered hearing a sharp crack in his ears—probably the snapping of one or more of the small bones in the middle ear—followed by a stabbing pain. And then, gathering silence.

"I haven't heard a bird sing since I was twelve years old," Edison would later say, without a trace of self-pity. Edison's hearing grew progressively worse, the damage probably exacerbated by his childhood bout with scarlet fever. Eventually, Edison would settle into a life of being nearly deaf. What hearing he did retain was oddly selective. Edison could make out shouted words and loud percussive sounds like the clack of a telegraph key, but the world of normal conversation was lost to him forever.

"I can hear talk in noisy places without much difficulty," Edison later said. "When I traveled between New York and Orange on suburban trains, while the train was running at full speed and roaring its loudest, I would hear women telling secrets to one another, taking advantage of the noise. But during stops, while those near to me conversed in ordinary tones, I could not hear a single word."

Edison acknowledged that his deafness fundamentally changed his life, but he would always insist that it was solely a change for the better. "This deafness has been of great advantage to me in various ways," he wrote. "When in a telegraph office, I could only hear the instrument directly on the table at which I sat, and unlike the other operators, I was not bothered by the other instruments. . . . My deafness has never prevented me from making money in a single instance. It has helped me many times. It has been an asset to me always."

Deafness drove Edison to reading, the young boy finding solace in the stacks of the Detroit Public Library. "I started with the first book on the bottom shelf and went through the lot, one by one," he recalled. "I didn't read a few books. I read the library."

Edison inhaled thick popular reference books like *The Penny Library Encyclopedia* and burrowed through serious tomes like Robert Burton's *Anatomy of Melancholy*, Edward Gibbon's *Decline and Fall of the Roman Empire*, and Isaac Newton's *Principia*. He was utterly baffled by Newton's work, finding the abstruse calculations incomprehensible. "I kept at mathematics till I got a distaste for it," Edison would later say.

Edison would credit his deafness with being instrumental in developing some of his greatest inventions. He spent long hours devising an improved telephone transmitter because he couldn't hear the sounds produced by the first Bell phones. He slaved over endless tweaks in the design of the phonograph because he couldn't make out recorded sounds that had any harsh overtones or hissing consonants. Later in life, when an ear specialist offered to perform an operation that would improve the inventor's hearing, Edison brusquely waved the man away.

"I wouldn't let him try," Edison said. "I know men who worry about being deaf although they are not half as deaf as I am. If they would let their deafness drive them to good books, they would find the world a very pleasant place."

But Edison's loss of hearing changed his world in ways that the great inventor didn't fully appreciate. Unable to make out much of what other people were saying, Edison eventually stopped trying to listen. Most of his conversations were more like monologues, especially after he became a famous inventor in his early thirties. Edison argued his points with great passion, at times raising his thin, reedy voice and pounding on a table to drive his case home. But when it came to opposing points of view, Edison literally didn't want to hear them. In an ordinary man, such behavior would be considered boorish; for an inventor, it would prove to be downright dangerous.

"The things that I have needed to hear," Edison would say, "I have heard."

Cut off from the world of sound, Edison's perspective became intensely visual. He made detailed sketches of many of his great inventions before creating them in the lab. His laboratory notebooks are filled with detailed drawings that look like the work of a draughtsman or an architect rather than the scribbles of an inventor. For Edison, inventions weren't real until he could see them in his mind and set them down on paper. He had little talent for abstraction and no patience for mathematics. His orientation was visual and linear, an approach that served him well as long as the problem was similarly constructed.

Electricity would prove to be a particularly difficult line for a visual man like Edison to pursue. Electricity was invisible, mysterious, elusive. You couldn't sketch electricity in a notebook or construct a faithful model of current zipping through a wire. Electricity required abstract thinking, putting Edison at a decided disadvantage when men began to devise ways of sending those elusive electrons across great distances to power the world. When he really needed to listen to other ideas about electricity, he would be locked in a silent world of his own making, hearing only his own opinions and the steady thrum of blood coursing through his ears.

"The things that I have needed to hear, I have heard."

Not long after his accident, Edison resumed his work on the Grand Trunk Railway, and soon came up with a way to supplement his income. He purchased a small portable printing press in Detroit and began cranking out a weekly newspaper devoted to news of the Grand Trunk line. "The Weekly Herald, Published by A. Edison" cost three cents a copy. Edison was not only the paper's publisher but also its compositor, pressman, news dealer, advertising rep, and sole reporter.

Only a few copies of the paper survive, but they clearly show a lad with a nose for news and little interest in the finer points of spelling or grammar. The paper contained telegraphic bursts from life along the rails: short items about births, lost and found parcels, business along the rail line, stagecoach departures. There were also several longer stories, written in the hard-boiled style of a big-city newspaper reporter. One of Edison's stories tells of an agent for the Haitian government who tried to swindle the Grand Trunk Railway company out of $67, the cost of a "valice" he was said to have lost, but was foiled "by the indomitable perseverance and energy of Mr. W. Smith, detective of the company."

Another article is full of praise for E.L. Northrup, one of the regular engineers on the Grand Trunk. Edison wrote: "We do not believe you could fall in with another Engineer, more careful, or attentive to his Engine, being the most steady driver that we have ever rode behind (and we consider ourselves some judge, having been Railway riding for over two years constantly) always kind, and obliging, and ever at his post." The glowing account reveals the young Edison already using the power of the press to his advantage. Engineers had ultimate authority over what was permitted on their trains; a few kind words in his newspaper could make an engineer look the other way when confronted with a mail car filled with oozing wet cell batteries and tiny bottles of chemicals marked "POISON."

Edison eventually lost interest in the newspaper, turning his restless attention to a device that had long held his fascination—the telegraph. Edison fashioned his own local telegraph system, stringing wire from the train station at Port Huron to the town center a mile away. He left his railroad job and became an apprentice telegraph operator in town, mastering Morse code and dutifully transcribing the nightly press reports off the humming wire.

As an apprentice telegrapher, Edison worked the overnight shift, and the tedium of spending long hours alone led him to create his first invention. During the night, telegraph operators were required to signal the number "6" to the train dispatch office every hour to prove that they were still awake at the key. Edison fashioned a small metal wheel with notches cut along the rim and attached it to a running clock. Each hour, the wheel spun around and tripped off a relay that automatically flashed the "6" message to the dispatch office, leaving Edison free to pursue his own experiments.

Good telegraph operators were hard to come by, and a skilled Morse code man could walk into almost any town and get a job on the spot. At age sixteen, Edison left home and began a five-year odyssey as a vagabond telegraph operator, taking briefly held jobs in Detroit, New Orleans, Cincinnati, Indianapolis, and Memphis. For a lad with only a few months' formal education, the telegraph was a godsend, opening vistas that would have otherwise been closed to him. The life of a journeyman telegraph operator was eye-opening for the small-town-raised Edison. He traveled extensively in the South shortly after the end of the Civil War, and found the towns of the defeated Confederacy quite an education.

"Everything was free," Edison recalled. "There were over 20 keno rooms running. One of them I visited was in a Baptist Church, the man with the wheel being in the pulpit and the gamblers in the pews."

At nineteen, he lingered for a time in Louisville, working as a telegrapher for Western Union. The Louisville telegraph office was in deplorable shape. Plaster flaked off the ceiling; the switchboard for the telegraph wires featured brass connections that were black with corrosion and neglect. Occasionally, the connection would short-circuit, resulting in a tremendous boom that Edison likened to a cannon shot. One room was filled with crumbling record books and stacks of message bundles, along with a hundred nitric-acid batteries arranged on a stand. The stand, as well as the floor beneath it, was all but eaten through by the acid dripping from the batteries. Despite the conditions, Edison thoroughly enjoyed the freedom such jobs gave him. He would always have a soft spot for the telegraph, long after the device was eclipsed by some of his own inventions. He would propose to his second wife, Mina, in Morse code, and his two children were nicknamed "Dot" and "Dash."

As an itinerant telegraph operator, Edison kept mostly to himself. "The boys did not take to him cheerfully, and he was lonesome," recalled one of Edison's coworkers. Edison had few friends and preferred to spend his spare time reading and tinkering with experiments. His deafness no doubt contributed to his splendid isolation.

"I was shut off from that particular kind of social intercourse which is small talk," Edison recalled. "I am glad of it. I couldn't hear, for instance, the conversations at the dinner tables of the boarding houses and hotels where after I became a telegrapher I took my meals. Freedom from such talk gave me an opportunity to think out my problems."

Edison's problems were largely scientific. Telegraph offices contained a treasure trove of supplies for the experimenter—old batteries, bales of copper wire, scraps of metal, and small hand tools. Using spare parts salvaged at work, Edison constructed a device that repeated the dots and dashes coming over the telegraph line at a slower speed and recorded them as indentations on a disk of paper. Operators could then "record" telegraph messages on the paper disk and transcribe them at their leisure.

Edison was keenly aware of the dual nature of electricity, positive and negative, creative and destructive. While working as a telegraph operator in Cincinnati, Edison got hold of a secondhand induction coil, the electrical transformer developed by Michael Faraday that could take a weak current from a chemical battery and increase it to a higher voltage. Edison rigged up the coil to a metal wash tank in the telegraph office, and then drilled a peephole in the ceiling through the roof. He invited several coworkers up to the roof to watch the spectacle unfold below.

"The first man entered and dipped his hands in the water," Edison recalls. "The floor being wet he formed a circuit, and up went his hands. He tried it the second time, with the same result. . . . We enjoyed the sport immensely."

Edison came up with another electrical device to deal with the rodents that infested the telegraph office, an invention that he dubbed the Rat Paralyzer. It consisted of two metal plates insulated from each other and connected with a main battery. The plates were placed so that when a rat passed over them, with its front feet touching one plate and its back feet the second, an electrical circuit was completed. At that instant, there was a brilliant flash of light, a loud pop, and one dead rat. Edison would later build a similar device to electrocute cockroaches.

Like Franklin and other experimenters, Edison occasionally found himself on the wrong side of electricity's dual nature. One day, while experimenting with an induction coil, he absent-mindedly grabbed both electrodes of the coil. A surge of electricity raced down his arms and contracted the muscles of his hands, further tightening his clutch on the electrodes. Edison had subjected dozens of creatures to the killing power of electricity; now he was caught in its death grip, frozen on the circuit.

"The only way I could get free was to back off and pull the coil, so that the battery wires would pull the cells off the shelf and thus break the circuit," Edison recalls. "I shut my eyes and pulled, but the nitric acid splashed all over my face and ran down my back. I rushed to a sink, which was only half big enough, and got in as well as I could and wiggled around for several minutes to permit the water to dilute the acid and stop the pain. My face and back were streaked with yellow; the skin was thoroughly oxidized. I did not go on the street by daylight for two weeks, as the appearance of my face was dreadful."

In 1868, Edison took a telegraph job in Boston, the country's leading center of invention, the cradle of Yankee ingenuity. It was in Boston that Edison committed himself to being a full-time inventor, inspired by the works of Michael Faraday. Edison stumbled upon a complete set of Faraday's works in a bookstore, and was captivated not only by the Englishman's novel electrical experiments but also by his worldview, which was remarkably similar to Edison's. The unschooled Faraday was indifferent to money and had only a rudimentary grasp of mathematics, but he rose to become a giant in the nascent field of electricity. Science wasn't for ivory tower theorists, Faraday believed, but for practical men willing to roll up their sleeves and discover nature's secrets in the laboratory, a view perfectly in line with Edison's.

Edison came home from his job at the telegraph office and pored over Faraday's works long into the night, sometimes up until breakfast the next morning. The books described many of Faraday's experiments in great detail, and Edison set about imitating the master.

"I think I must have tried about everything in those books," Edison recalled. "[Faraday's] explanations were simple. He used no mathematics. He was the Master Experimenter. I don't think there were many copies of Faraday's works sold in those days. The only people who did anything in electricity were the telegraphers."

Edison had always had a passion for tinkering, but it was more something he did in his spare time when the boss was looking the other way. In Faraday, he saw another possibility: the life of a full-time inventor. In January 1869, a small item in a trade journal announced his intentions to the world: Thomas A. Edison, formerly a telegraph operator, "would hereafter devote his full time to bringing out his own inventions."

Once Edison knew what he wanted to do with his life, the inventions began to pour out of him like water. He was granted 38 patents in 1872 and another 25 the following year, many of them having to do with improvements in the telegraph. The inventions kept coming; at the height of his powers, Edison was granted 106 patents in a single year. Before his inventing days were through, Edison would be granted a staggering 1,093 patents in the United States alone.

Edison's first-ever registered invention, patent number 90,646, was granted on June 1, 1869. It was for an Electric Vote-Recorder, a telegraph-like device that electronically tabulated votes cast in legislatures. It was an elegant piece of engineering—but an utter failure as an invention. Most legislators of the day didn't want their votes counted quickly, preferring to withhold their ballots as long as possible in order to make speeches and cut backroom deals. Edison failed to sell his vote recorder to the U.S. Congress, and his first invention would amount to little more than a sketch gathering dust in the Patent Office. Edison took the lesson to heart. His later inventions would demonstrate not only breakthrough technology but also a keen eye for what the public wanted and would accept.

After the vote recorder, Edison's next invention fared far better—an improved stock market ticker that earned him $40,000 virtually overnight. The inventions followed thick and fast through the early 1870s—an electric pen, an improved battery, paraffin paper (used to wrap candies), a duplex telegraph (which could send two independent messages over the same wire), as well as a quadraplex (four independent messages) and sextuplex (six messages) model. He also invented the carbon-button transmitter, a version of which is still used today in microphones and telephone receivers.

Edison was so prolific that he didn't have time to develop all his inventions. In 1875, he came up with a device for making multiple copies of letters, which he dubbed the mimeograph. With other inventions already taking up his attention, Edison sold the rights to the mimeograph to A.B. Dick, a Chicago firm that would become one of the world's leading office supply companies. Generations of students would long savor the distinctive smell of mimeographed test papers, without realizing they were getting a whiff of another Edison invention.

Only someone with Edison's creative genius could have come up with so many new inventions and refinements of existing devices. But Edison was also a man of his time, born at the right moment in history to apply his considerable powers to practical science. Edison came into adulthood at the dawn of the Industrial Age. Had he been born twenty years earlier, he would have found few opportunities as an inventor; had he come along twenty years later, he might have ended up a frustrated researcher at one of the large industrial corporations. Edison was at the right place at the right time with the right mind.

In 1876, Edison built a state-of-the-art "invention factory" where he could continue his work. He set up shop in Menlo Park, New Jersey, about twenty miles outside New York City, constructing what could be considered the first modern research and development center in the world. The Menlo Park laboratory employed dozens of workers, and later hundreds, all toiling on various Edison projects. His men soon learned to adapt themselves to their boss's trial-and-error methods. As one of his workers recalls, "Edison seemed pleased when he used to run up against a serious difficulty. It would seem to stiffen his backbone and make him more prolific of new ideas."

Edison would pursue many different inventions and lines of inquiry at the same time; only in a few instances would a single invention consume his attention. One such invention was the phonograph, which dominated his attention throughout 1877.

Edison had been tinkering with an automatic method of recording telegraph messages on a disk when he realized such a machine might be put to even better use. He built a device consisting of a large cylinder wrapped in tinfoil, which engaged a small chisel-like recording stylus, which in turn was connected to the center of an iron diaphragm. By rotating the cylinder and then speaking into the diaphragm, he caused the needle to vibrate and make a series of indentations in the foil corresponding to the sound waves. For playback, a second stylus traced the indentations in the foil.

Edison didn't expect much from the device—at best he thought he might be able to record a barely discernable word or two. Edison cranked the cylinder and shouted a verse of "Mary Had a Little Lamb" into the diaphragm. He placed the playback needle at the start of the foil indentations, cranked the cylinder once again and was amazed to hear his own voice talking back to him.

"I was never so taken aback in my life," Edison recalled. "Everybody was astonished. I was always afraid of things that worked the first time."

The phonograph would be Edison's favorite invention, mainly because nothing quite like it had been created before. His determination in perfecting the device was relentless. To test the sound, he would place his ear directly on the phonograph horn. Sometimes he'd even bite into the horn with his teeth, letting the sound vibrations ring through the bones in his head. For a deaf man, inventing a talking machine was almost a miracle.

"The phonograph never would have been what it now is if I had not been deaf," Edison said. "Being deaf, my knowledge of sounds had been developed till it was extensive and I knew that I was not getting overtones. . . . It took me twenty years to make a perfect record of piano music because it is full of overtones. I now can do it—just because I'm deaf."

Writing in 1878 in the *North American Review*, Edison shared his ideas about the future applications of his new-fangled invention:

> Among the many uses to which the phonograph will be applied are the following:
>
> 1. Letter writing and all kinds of dictation without the aid of a stenographer.
> 2. Phonographic books, which will speak to blind people without effort on their part.
> 3. The teaching of elocution.
> 4. Reproduction of music.
> 5. The "Family Record"—a registry of sayings, reminiscences, etc., by members of a family in their own voices, and of the last words of dying persons.

6. Music-boxes and toys.
7. Clocks that should announce in articulate speech the time for going home, going to meals, etc.
8. The preservation of languages by exact reproduction of the manner of pronouncing.
9. Educational purposes; such as preserving the explanations made by a teacher, so that the pupil can refer to them at any moment, and spelling or other lessons placed upon the phonograph for convenience in committing to memory.
10. Connection with the telephone, so as to make that instrument an auxiliary in the transmission of permanent and invaluable records, instead of being the recipient of momentary and fleeting communication.

Most of Edison's predicted applications for the phonograph had to do with recorded speech; the eventual hands-down winner, number 4, the reproduction of music, was tucked away as an afterthought. Edison would always have a sharp eye for new inventions, but he had a tin ear when it came to predicting how people would use his devices. Edison predicted that the motion picture camera would one day be a great tool for education, with film eventually supplanting books in schools and universities. And electricity, he believed, would one day be generated by small power stations located in every town, and consumed by homes and businesses close to the plant.

When Edison announced his phonograph to the world, he became an overnight celebrity. The device seemed magical to the public, and gentleman journalists from the big newspapers scurried out to interview him about his wonderful invention. "Such an invention will be of inconceivable practical good to business men and to public speakers," declared the *Boston Times*. "Edison is giving to mankind far more than will ever be returned to him under any patent he may ever take out." The press dubbed him "The Wizard of Menlo Park." It was a role Edison had been rehearsing for most of his life.

4

LET THERE BE LIGHT

The phonograph was an unexpected invention, a lightning bolt of ingenuity that happened to touch down on earth. It attracted worldwide acclaim for its remarkable synthesis of science and art: a machine that could talk, invented by a man who was practically deaf. For Edison, though, there was something vaguely unsatisfying about the phonograph. Many considered it to be little more than a toy, a parlor amusement that would enjoy a brief flash of popularity and then quickly fade from view. Indeed, this opinion turned out to be largely correct for a time; the phonograph wouldn't enjoy widespread popularity for another twenty years.

Edison was a man who, above all, took invention seriously. It was his calling, his life, the window through which he viewed the world and himself. He soon began casting about for another, more weighty invention, one that everyone would immediately recognize as being essential to daily life.

Edison found a clue to his next invention in a department store in Philadelphia. In 1878, John Wanamaker's, one of the country's first department stores, became the first business to install arc lamps on its retail floor. Arc lamps were the predecessors of incandescent lights, producing illumination by the sparking (or arcing) of high current between two carbon electrodes. The twenty Wanamaker arc lamps gave off a brilliant light, bathing the merchandise in an almost blinding glow. A large circular counter stood at the center of the store, with more than a hundred counters of goods radiating around it, all glowing with artificial light.

Arc lighting was harsh, more suited for a prison yard than a department store. But the Wanamaker lamps created an immediate sensation. Thousands of people came to the store just to marvel at the lighting, the first major indoor installation of electrical lamps in the country. One visitor rhapsodized that the Wanamaker lights looked like "twenty miniature moons on carbon points held captive in glass globes." Many came away from the store convinced they had glimpsed the future.

Edison was more impressed by the oversized reactions the Wanamaker lights produced than by the technology of the lamps themselves. Arc lamps worked by passing an electrical current between two narrowly separated carbon rods. The problem was, the lamps were smoky and notoriously unreliable. The arc, or gap between the two carbon rods, burned at thousands of degrees, heating the carbon tips until they glowed. As the tips burned down, the gap between the rods had to be continually adjusted to keep the light burning, requiring almost daily care. Even then, the carbon tips wasted away quickly.

The Wanamaker lamps, produced by the C.F. Brush Company, represented the state of the art in electric lighting at the time. But Edison soon came to feel that the entire arc lighting approach was wrong. There was another way to create artificial illumination with electricity—incandescent lighting—but that technology was beset with even more problems than arc lighting. In 1860, back when Edison was hawking newspapers on the Grand Trunk Railroad, English physicist and chemist Joseph Swan invented what is generally considered to be the first incandescent lamp. In Swan's primitive lamp, an electric current was applied to a filament of carbonized paper encased in an airtight glass bulb. The filament, largely resistant to the electricity flowing through it, converted the electrical energy into heat, which in turn made the filament glow, or become incandescent. The problem with Swan's incandescent lamp, and others that followed, was that the filament quickly turned to ash under the intense heat.

Little progress had been made with incandescent lighting in the nearly twenty years following Swan's invention, and many believed that the incandescent approach was a technological dead end. After all, the ideal filament for an incandescent bulb had to have the qualities of the Biblical bush that spoke to Moses: it had to burn but not be consumed. It was just the sort of impossible challenge Edison loved.

From the start, Edison set out to invent not simply an incandescent lamp but an entire electrical system that would power the lamps along with future electrical inventions. Edison quickly realized that his main competitors weren't the handful of arc light manufacturers but rather the gas companies, which dominated the lighting market at the time. To build a successful rival to gas, Edison would have to come up with a way to "subdivide" direct current to power individual lamps, just as the gas companies apportioned small units of natural gas to customers. Edison's first step, in the summer of 1878, was to learn everything he could about the gas lighting market.

"I had made a number of experiments on electric lighting a year before this," Edison later recalled. "They had been laid aside for the phonograph. I determined to take up the search again and continue it. On my return home I started my usual course of collecting every kind of data about gas; bought all the transactions of the gas-engineering societies, etc., all the back volumes of gas journals, etc. Having obtained all the data, and investigated gas-jet distribution in New York by actual observations, I made up my mind that the problem of the subdivision of the electric current could be solved and made commercial."

Edison's lab notebooks, most of which have survived, are crammed with notations referring to "Electricity vs. Gas as General Illuminants." One early entry sets out the grand mission: "Object, Edison to effect exact imitation of all done by gas, so as to replace lighting by gas by lighting by electricity. To improve the illumination to such an extent as to meet all requirements of natural, artificial, and commercial conditions." From the start, Edison believed that in the future, gas would be used less for lighting and more for heating, with electricity taking over the lighting chores. Edison made copious notes about the entire range of devices he'd need to fashion to create a complete electrical utility to compete with gas, down to the meters he'd need to measure consumption and bill customers. He noted the weaknesses of gas lighting, in anticipation of a marketing campaign he'd launch to convince customers to switch from gas to electric: "So unpleasant is the effect of the products of gas that in the new Madison Square Theatre every gas jet is ventilated by special tubes to carry away the products of combustion," Edison wrote. It wasn't long before Edison knew more about the gas industry than most people in the business.

While Edison was mapping out his business strategy, he struggled to make headway in designing a reliable incandescent bulb. He tested the electrical resistance of hundreds of incandescent materials, and fashioned a crude prototype lamp with a filament made of platinum. The bulb burned for all of ten minutes before going out, hardly the sort of performance that would convince gas customers to switch to electric light.

Edison wasn't discouraged by his slow progress. In fact, it only seemed to embolden him. In the fall of 1878, with little more than his ten-minute bulb in hand, Edison began calling on reporters from the New York newspapers, drumming up publicity for his latest invention. The Wizard knew how to make good copy—after all, he had seen the newspaper game from the inside during his days on the Grand Trunk Railroad—and journalists were happy to take the train out to Menlo Park to interview such a reliably colorful subject. Edison wanted the publicity not to satisfy his own vanity, but rather to attract Wall Street investors. The hunt for a universal incandescent bulb was going to be an expensive venture, and Edison would need substantial backing for research and development.

In interviews, Edison boldly declared that he had "already discovered" a system for turning electricity into a cheap and practical substitute for gas; it was just a matter of working out the bugs. He held several demonstrations for reporters, who, unschooled in the ways of electricity, took Edison's interpretation of the tests at face value. When Edison demonstrated his platinum wire bulb for a reporter for the *New York Sun*, the journalist could scarcely believe his eyes:

"The new light came on, cold and beautiful," the reporter wrote. "The strip of platinum that acted as a burner did not burn. It was incandescent . . . it glowed with the phosphorescent effulgence of the star Altair. A turn of the screw and . . . the intense brightness was gone."

The exhibition was impressive, but that was because Edison was at the controls. Had Edison not shut off the current flowing through the platinum wire almost immediately, the filament would have quickly burned out, taking with it the phosphorescent effulgence of the star Altair. Instead, the brief demonstration allowed Edison to make even bolder claims—his lamp meant the death of the gas industry.

"When ten lights have been produced by a single electric machine, it has been thought to be a great triumph of scientific skill," Edison told reporters. "With the process I have just discovered, I can produce 1,000, ay, 10,000 from one machine. Indeed the number may be said to be infinite. When the brilliancy and cheapness of the lights are made known to the public—which will be in a few weeks, or just as soon as I can thoroughly protect the process—illumination by carbureted hydrogen gas will be discarded."

Edison's press campaign had its desired effect; the moneymen of Wall Street rushed to get in on the action. A consortium of Wall Street financiers, including W.H. Vanderbilt (then the country's richest man), J.P. Morgan, and the directors of Western Union (Edison's one-time employer) put up a total of $300,000 to create a new company, the Edison Electric Light Company. Edison received the money in installments to fund his experiments at Menlo Park; in return he agreed to assign to the newly formed company all his inventions in the lighting field for the next five years.

With funding secured, Edison had everything he needed to begin producing incandescent lamps—everything, that is, but the design of the lamp itself. Edison's initial prediction of producing a reliable, long-lasting incandescent bulb "in a few weeks" would prove to be wildly optimistic, sorely testing the inventor's spirit.

Edison plunged into the task of improving his platinum wire lamp with a world-class laboratory at his disposal. His Menlo Park workshop steadily expanded, growing to a staff of as many as sixty machinists, carpenters, and lab workers. Most of the serious work took place on the upper floor of the hundred-foot-long laboratory building, a cavernous hall containing several long wooden tables, which were usually strewn with scientific instruments and notebooks. In the rear of the hall sat a pipe organ, which the near-deaf Edison would occasionally play, as one listener put it, "in a primitive way." At dusk, when the rays of the setting sun filtered through the windows casting moody shadows, the hall looked like the laboratory of a mad scientist.

Next to the main hall was the machine shop, a large brick building that contained lathes, drilling machines, and tools. Almost any apparatus could be constructed in the machine shop, built to Edison's exact specifications. Since Edison always preferred to "see" his inventions before actually fashioning them, he installed a cyanotype machine, an early version of a blueprint maker, which he used for making multiple copies of drawings and plans. There was also a carpenter shop, a glass-blowing shed (to fashion light bulbs) as well as a gasoline plant that powered the complex. While working on the incandescent lamp, Edison and his men were illuminated by gas lighting.

Edison and his assistants put in long hours; Edison himself usually slept only four hours a night. He was deadly serious about his work, although he could always be interrupted for a good joke. One associate described Edison's laugh as "sometimes almost aboriginal." After hearing a funny story, the inventor would slap his hands delightedly on his knees and rock back and forth with pleasure. Edison seldom drank alcohol, though he enjoyed smoking cigars. He often held a thick black cigar tightly in his mouth while working, and he eventually developed a "Havana curl," a slightly deformed upper lip, from his constant smoking. Edison's only real vice was working too much. His wife, Mary, who married Edison in 1871, accepted the fact that she had to share her husband with a lifelong mistress, invention.

Two months passed after Edison's brassy promise to produce a working incandescent bulb in a matter of weeks, and some began to wonder whether Edison had overplayed his hand. The *New York Herald* dispatched a reporter to Edison's lab in mid-December 1878, to see what was delaying the production of the incandescent lamp. The reporter found Edison "seated at a long wooden table, on which were promiscuously scattered a dozen or more scientific books on heat, light, and electricity, eight or ten cells of a battery . . . and two of his instruments for the testing of the new electric light. The inventor was sitting with his chin resting on his hands, his elbows on the table."

When the *Herald* man got around to asking Edison about the progress of his electric light, the inventor was as optimistic as ever.

"It is all completed now, so far as the principle is concerned," Edison said confidently. "It is now only a question as to cost, but one thing you can say—it is established beyond doubt that it is cheaper than gas. It is a better and cheaper light."

But Edison still had nothing to show the public. The inventor's rivals seized on the delay as proof that the incandescent lamp existed only in Edison's imagination. Gas company executives, who had seen their stocks drop more than 10 percent in value since Edison's publicity campaign, dismissed the inventor as a flim-flam artist whose talk of cheap incandescent light was little more than a ruse to shake money out of investors. Some scientists insisted that electrical current could not be "subdivided" as Edison claimed, and that the inventor's approach was scientifically unsound.

Edison was feeling the heat. "It must be confessed," Edison wrote at the time, "that hitherto the 'weight of scientific opinion' has inclined decidedly toward declaring the [incandescent light] system a failure, an impracticality, and based on fallacies. It will not be deemed discourteous if we remind these critics that scientific men of equal eminence pronounced ocean steam-navigation, submarine telegraphy, and duplex telegraphy, impossible down to the day when they were demonstrated to be facts." The delay in coming up with a practical incandescent light system was due, Edison said, "to the enormous mass of details which have to be mastered before the system can go into operation on a grand scale."

Edison wasn't simply developing a workable incandescent light; he was designing an entire commercial system that would produce and distribute direct current, creating a technical standard from scratch. The incandescent lamp was only a piece of the puzzle, albeit an essential one. Without the lamp, an electrical distribution system was useless.

Developing the incandescent light, Edison ran into two critical hurdles. One was finding a way to exhaust enough air out of the glass bulb to produce a near-perfect vacuum. If even one ten-thousandth of an atmosphere of air remained in the bulb, the oxygen weakened the filament. After trying a variety of hand pumps and being unsatisfied with the results, Edison turned to a recently invented device from England, the Sprengle pump, which used mercury to trap air bubbles in the bulb and expel them. Edison obtained one of the first Sprengle pumps in America and immediately put it to work. He and his assistants furiously pumped glass bulbs for hours and found that the Sprengle pump produced a vacuum that came within one or two millimeters of complete air exhaustion. It was a small but crucial breakthrough.

The second challenge was finding the right filament. It would have to resist a tremendous amount of heat—more than 1,000 degrees Fahrenheit just to get a feeble red glow—without being consumed, all the while emitting a steady, unflickering light. The search for the perfect filament was the sort of blind treasure hunt Edison loved. He tried using carbon because of its high melting point, but found it burned up quickly. Platinum had shown promise in his prototype lamp, but that too had a short life, and was expensive. Edison's laboratory was crammed with rare and exotic metals from around the world, and he tried nearly every one of them as a filament: chromium, boron, osmium, platinoiridium, molybdenum. (Edison considered tungsten, the highly heat-resistant metal used in many modern-day incandescent bulbs, but lacked the tools to handle the element properly.) Whenever a material exhibited even the slightest promise, Edison jotted it down in his laboratory notebook, along with the notation "T.A." for Try Again.

The filament tests were especially taxing, forcing Edison to stare at blinding white-hot metal for hours on end. Edison's notebook entry for January 27, 1879, reports: "Owing to the enormous power of the light my eyes commenced to pain after seven hours work and I had to quit." The next day was like a bad hangover: "Suffered the pains of hell with my eyes last night from 10 p.m. to 4 a.m. when got to sleep with a big dose of morphine." Later the same day, after the morphine buzz wore off, Edison wrote: "Eyes getting better and do not pain much at 4 p.m., but I lose today."

Thanks to the Sprengle pump, the near-perfect vacuum in the glass bulb was much more forgiving on filaments brought to incandescence. Edison returned to testing platinum, and found that in a high vacuum, a coil of platinum became extremely hard while remaining resistant to electricity, giving off a pleasing orange glow when it reached incandescence. Platinum was costly, but Edison felt he could make the filaments thin enough so they'd contain only a small amount of the metal. On April 12, 1879, Edison executed a patent on the high-resistance platinum lamp featuring the improved vacuum. The search for the universal incandescent bulb, Edison declared, was over. His platinum filament lamp would bring light into the darkest corners of the earth, at a cost less than half that of gas. The *New York Herald* declared: "THE TRIUMPH OF THE ELECTRIC LIGHT."

Once again, Edison's hopes ran ahead of reality. There were problems with the new platinum lamps almost as soon as they were subjected to less than ideal laboratory conditions. The platinum coils consumed a great deal of power for the amount of light they gave off, and the bulbs weren't very reliable when the current was increased. Shortly after his patent application, Edison gave a private demonstration of platinum filament lamps for a cadre of investors. The results were downright embarrassing; most of the bulbs popped within a few seconds of reaching incandescence. Edison's investors grumbled in the darkness and left more worried than ever.

Word of the lamp's poor performance leaked out to the newspapers; after all, there were plenty of people in the gas industry whose livelihoods depended on Edison's failure. Shares of the Edison Electric Light Company fell sharply in price and stories critical of Edison appeared in some of the same newspapers that had praised him just months before. One account declared, "It is now known that Mr. Edison has failed in his experiments. . . . The inventor has never been able to regulate his current so as to keep his lamps burning for any length of time, and he has never ventured on a single public exhibition of it." Edison finally realized that all his talk about the incandescent light was doing more harm than good. Against his natural inclinations, he began dodging the press.

"You will have to excuse me," Edison wired one reporter (for he was still a telegrapher at heart). "Talking or writing about the electric light won't make it a success. The moment the light is finished, the public shall have it. Before it is finished I would rather not talk about it."

Edison's investors were already restless over his lack of progress on the lamp. Now his decision to stop talking to the newspapers meant that his detractors had free rein to raise doubts about the project. W.H. Preece, the chief electrician for the British Post Office, declared that the electric light was no match for gas. Electric light "does not lend itself to distribution like the gas flame," Preece said, and was further held back by "the unsteadiness of the light due to variations in the speed of the engine employed in driving the dynamo machine." The subdivision of light was, Preece declared, "an absolute *ignis fatuus*," literally, a "foolish fire."

The rumble of opposition made Edison more determined than ever to succeed; he loved nothing better than proving the so-called experts wrong. Edison often needed to have his back against the wall before he could move forward. As Edison's chief scientific assistant at Menlo Park, Francis Upton, put it, "I have often felt that Mr. Edison got himself purposely into trouble by premature publications and otherwise, so that he would have a full incentive to get himself out of the trouble."

Back in his laboratory, Edison returned to first principles. Above all, the lamp would need to be fed with a steady, reliable flow of direct current. For the purpose, Edison built from scratch an improved dynamo; a tall upright mass of iron, magnets, and coiled wire nicknamed "Long-Waisted Mary Ann." The dynamo demonstrated an astonishing 90 percent efficiency in converting steam to electricity; the best dynamos on the market had only a 40 percent efficiency. Edison experimented with various voltages, or electrical pressures, to supply to his incandescent lamp. Too much voltage and the delicate filament would quickly overheat and break; not enough and the light would flicker. Edison finally settled on supplying his lamps with 110 volts of electricity, a decision that, more or less, is still with us today. The United States, Canada, Mexico, Japan, and a handful of other countries operate on a 110- to 120-volt electrical system.

Finding the right filament remained the critical problem. Edison found some success with elements that had been carbonized, that is, baked in a high-temperature oven until there was little left except the compound's essential carbon framework. Edison's notebooks filled with still more filament candidates, all subjected to carbonization: cardboard, drawing-paper, paper saturated with tar, fishing line, thread rubbed with tarred lampblack, cotton soaked in boiling tar, coconut hair. About 1,600 different materials were tested until Edison finally hit upon an unlikely winner—a strand of ordinary cotton thread. Edison placed the thread in an earthenware mold, baked it under high heat in an oven, and then carefully removed the carbonized thread from the mold. The thread was everything Edison had been looking for: strong, whisper thin, highly resistant to electricity, and able to withstand intense heat without breaking.

Edison tested the carbonized cotton thread extensively during marathon lab tests on October 21 and 22, 1879, red-letter dates in the history of electricity. Edison's notebook reads, "October 21—No. 9 ordinary thread Coats Co. cord No. 29, came up to one half candle and was put on 18 cells battery permanently at 1:30 A.M." The carbonized piece of ordinary thread tested on lamp number 9 glowed to incandescence for 13½ hours before breaking, by far the longest-lasting filament Edison had produced. Eventually, Edison found an even more durable material: a tough paper known as Bristol cardboard.

This time, Edison tried to keep mum until he was absolutely sure he had a working lamp to show to the public. He told the *New York Times*, "All the problems which have been puzzling me for the last 18 months have been solved," but the paper viewed his statements with some skepticism, headlining one story "EDISON'S ELECTRIC LIGHT: CONFLICTING STATEMENTS AS TO ITS UTILITY." The same story described Edison, somewhat uncharitably, as "a short, thick-set man with grimy hands."

On November 1, Edison executed a patent for a carbon filament lamp, which was granted as U.S. patent number 223,898. The glass globe was now rounded; the filament was shaped like a horseshoe. There was no light switch; the lamp was turned off by unscrewing the bulb. It wasn't the first electric light, or even the first incandescent lamp. It was, however, the first practical long-lasting incandescent bulb, the dawn of the age of electric light.

Edison kept the full story out of the newspapers for nearly two months until he approached the *New York Herald* and collaborated on a lengthy article. Published on December 21, 1879, the story was announced by a lead editorial bearing the headline "EDISON'S EUREKA—THE ELECTRIC LIGHT AT LAST." The article described Edison's lamp as giving off a light "like the mellow sunset of an Italian autumn." It went on to note that the light produced "no deleterious gases, no smoke, no offensive odors—a light without flame, without danger, requiring no matches to ignite, giving out but little heat, vitiating no air, and free from all flickering; a light that is a little globe of sunshine, a veritable Aladdin's lamp."

The news of Edison's lamp reverberated around the world. In the week following Christmas 1889, hundreds of visitors made a pilgrimage to Menlo Park to see the marvel for themselves, so many that the railroad had to run extra trains to Menlo Park. On New Year's Eve, the throng grew to several thousand, including a *New York Tribune* reporter, who described the scene: "By eight o'clock the laboratory was so crowded that it was almost impossible for the assistants to pass through. The exclamation, 'There is Edison!' invariably caused a rush that more than once threatened to break down the timbers of the building." Those who came to Menlo Park never forgot the sight of the glowing lamps, even if many didn't understand how they worked. More than one visitor asked Edison how he had gotten the red-hot horseshoe into the glass globe without burning his hands.

It was an astounding triumph for Edison, producing a working incandescent lamp in just fifteen months. It was a work of pure creative genius; forever after, the cultural shorthand for a bright idea would be a picture of a person with a light bulb over his head.

But for Edison, the incandescent lamp was only the first part of a much larger plan, one that would transform the world.

5

ELECTRIFYING THE BIG APPLE

"My light is perfected," Edison announced. "I'm now going into the practical production of it."

In February 1881, Edison moved from Menlo Park to New York City to fulfill his next mission: bringing electric power to the Big Apple. He and his staff moved into a four-story brownstone at 65 Fifth Avenue, just off Union Square. A neat black sign with gold letters greeted visitors at the door: "The Edison Electric Light Company."

Word that the Wizard had moved to New York set off a buzz even before Edison had strung a single wire. A throng of visitors descended on Edison's Fifth Avenue headquarters almost as soon as he moved in. "Visitors seven deep awaited their turn," one journalist reported, "while Edison, in a black morning coat, silk wrapping about his throat, and the invariable cigar, explains his work and his schemes with untiring repetition."

Edison was riding the crest of a remarkable wave during which he had invented the phonograph, perfected the incandescent lamp, and laid the groundwork for a complete electrical system to generate and transmit direct current to customers. There were fewer doubters now, as most of his brash predictions had come true, though not always exactly when and how Edison said they would. Edison was increasingly being addressed as "Professor Edison" despite his scant schooling and utter disdain for formal education. Some newspapers, uncertain what to call a man of Edison's expertise, referred to him as an "electro-scientist." Even the language was struggling to keep up with him.

Edison soon added teacher to his list of credits. The top floor of his Fifth Avenue headquarters was converted into a school for about thirty students, with Edison instructing the class on the practical fundamentals of electricity. There were no textbooks and few electrical standards besides the ones Edison himself had devised. Edison was literally making up the rules as he went along. Before he was done, Edison and his team would have to invent a complex system of interlocking technologies to complement the incandescent lamp: switches, meters, sockets, fixtures, regulators, underground conductors, junction boxes, and, most important, a central station to generate DC power and a distribution network to deliver it.

Edison agents began sniffing out potential clients through door-to-door surveys, asking people about their experiences with gas, or what the Edison men liked to call "the old-time light." Many customers complained about gas leaks, flickering lights, unpleasant odors, and unreliable pressure in the gas mains. The power market was ripe for a new player.

Edison knew he'd have to get his electrical system right the first time—the gas companies were sure to pounce on any misstep. If electricity hoped to replace gas as a source of light, it would have to be safe and above all, reliable. Most of the bugs would have to be worked out before the system went into place. Edison constructed a large working model of an electrical distribution system on the grounds of his Menlo Park laboratory. Eight miles of electrical wiring were buried in the ground, supplying power to six hundred lamps dotting the property. The model let Edison troubleshoot his distribution system before bringing it to market; engineers made economy tests to make sure the system was commercially viable. It also gave the visually oriented Edison a chance to "see" his system in action before installing it on a large scale.

The gas companies began attacking the Edison electrical system even before it was built. The *American Gas Light Journal*, a trade publication, sniffed that the incandescent lamp provided "illumination but not lighting," and warned of the light's "evil effects on the eyes." It went on to declare: "European observers state that the frequent variations in intensity to which the light is subject give rise to sudden and frequent changes in the pupil. Such a light, therefore, causes not only muscular fatigue but also a considerable degree of blurring and indistinctness in the retinal image."

The gas companies also pointed to electricity's hidden dangers, tapping into primal fears as old as lightning. The 110 volts of direct current that Edison planned to run into customers' homes and offices was enough to cause fire, injury, and even death. Edison had a healthy respect for electricity's dark side from the scores of shocks he had suffered during the course of his experimenting. To minimize the risk of fire and accidents, Edison decided to run his electrical lines underground, rather than on overhead poles, which were already bristling with telegraph and telephone wires.

To soothe customers' fears of the new technology, Edison made sure that his electrical system would resemble the existing gas system as closely as possible. The illumination of an electrical lamp was set at 16 candlepower, the same brightness as a gas light. The incandescent lamps could be turned on and off with a key, just like gas lamps. The Edison lamps were even referred to as "burners." At the same time, electricity had none of the problems associated with gas, which one Edison Electric advertisement enumerated in exhaustive detail for customers: "The disadvantages of gas are: sulphur thrown off, ammonia thrown off, air consumed, unsteadiness of light, danger from suffocation, danger from use of matches, expense from leaks in pipes, metals tarnished, carbonic acid thrown off, sulphurated hydrogen thrown off, atmosphere vitiated, colors made unnatural, excessive heat produced, danger from leaks in pipes, danger from fires, blackening of ceilings and decorations, freezing of pipes, water and air in pipes."

Edison's marketing message was clear: Electricity was just like gas, only better; it was new and improved. There was no flickering, no odor, and no danger of explosion. Electricity was portrayed as a "modern" power source in contrast to "old-fashioned" gas and coal, a marketing strategy that would continue well into the twentieth century, when the company Edison founded extolled the wonders of the all-electric kitchen.

While building his electrical system, Edison continued to tweak the incandescent lamp. He tested no fewer than six thousand vegetable growths as lamp filaments, and ransacked the globe in search of the ideal incandescing material. One of Edison's agents, schoolteacher James Ricalton, was sent on a yearlong treasure hunt through Asia trying to track down a rare bamboo fiber that had shown promise as a filament.

"I at once reported to Mr. Edison," Ricalton recalled upon his return to America, "whose manner of greeting my return was as characteristic of the man as his summary and matter-of-fact manner of my dispatch. His little catechism of curious inquiry was embraced in four small and intensely Anglo-Saxon words—with his usual pleasant smile he extended his hand and said: 'Did you get it?' "

Ricalton's bamboo fiber turned out to be another dead end, a failure that in Edison's mind only brought him slightly closer to the answer. Eventually, Edison settled on a bamboo fiber he discovered in a hand fan, which gave his lamps a life of about twelve hundred hours, compared to the ten to fifteen hours of a cardboard filament bulb. It was the first truly long-lasting incandescent lamp.

While experimenting with bamboo filaments, Edison noticed that once the filament had burned for several hours, carbon deposits blacked the inside of the bulb. It was a curious effect; the carbon moved through the bulb even though it was exhausted of air. Furthermore, the carbon seemed to be coming from the tip of the filament that was connected to the positive pole of the power supply, which implied that it was carrying an electrical charge. If so, that meant that electricity was not only flowing through the filament, but also through the vacuum inside the bulb—electricity without wires.

Edison had no explanation for the strange effect, even after he fashioned a bulb with a third electrode, which collected and measured the mysterious current. Edison patented the three-element bulb in November 1883, still unsure exactly what it could be used for, and soon moved on to other experiments.

Edison didn't understand the importance of his discovery at the time, but the curious flow of free electrons from an incandescent metal through a vacuum would come to be known as "the Edison effect." He had stumbled onto a device that would become a fundamental component of radios and televisions in the twentieth century: the vacuum tube. Such tubes were a mainstay of electronics until smaller, cheaper, and more durable transistors supplanted them. Like Benjamin Franklin, Edison believed that a discovery was valuable only if it was immediately practical, so he put the vacuum tube aside and returned to developing his electrical system.

The centerpiece of the system, the Edison lamp, was already a success—nearly forty thousand Edison lamps were sold the first year they were offered to the public. Edison cannily sold his lamps at a loss, more interested in dominating the electric light market than in turning a quick profit. In 1881, it cost Edison $1.10 to manufacture an incandescent lamp, but he sold the lamps for just 40 cents. The next year, he brought his manufacturing cost down to 70 cents, and still sold the lamps for 40 cents. By the fourth year of operation, Edison's manufacturing cost came down to 37 cents per lamp, and he made up all the money he had lost previously in just one year. It was a shrewd way to build a self-sustaining monopoly.

With the lamp well on its way, Edison concentrated on his next task: building an electrical power station in New York City. He hung a large map of Manhattan on a wall of his Fifth Avenue headquarters, which he surveyed like a general about to go into battle. Ideally, the plant would be located in an area close to businesses that would buy electricity, but where the land was still relatively cheap. Edison's initial foray into the New York City real estate market proved sobering for the small-town-raised inventor.

"I thought that by going down on a slum street near the waterfront I would get some pretty cheap property," Edison recalled. "So I picked out the worst dilapidated street there was, and found I could only get two buildings, each 25 feet front, one 100 feet deep and the other 85 feet deep. I thought about $10,000 each would cover it; but when I got the price I found that they wanted $75,000 for one and $80,000 for the other. Then I was compelled to change my plans and go upward in the air where real estate was cheap. I cleared out the building entirely to the walls and built my station of structural ironwork, running it up high."

Edison settled on a property located on Pearl Street, a grimy avenue in lower Manhattan, two blocks from the East River. In some respects, the area was an unlikely choice as the first to be electrified in the city. The area contained mostly office buildings and small factories and few residences, so most electricity would be consumed only during daylight hours. But this section of the city also included Wall Street, a powerful constituency that Edison was always mindful to court. Edison wanted to show investors that his electrical system had the potential for enormous profits down the road. Nothing would get the attention of overheated speculators faster than seeing the miracle of monopoly light up before their very eyes.

Edison snapped up the site at 255–257 Pearl Street in August 1881, and set to work on building the country's first electrical power plant from the ground up. The 110-volt DC system Edison had developed could transmit electricity about one square mile, limiting the area he could electrify from a single power plant. (In principle, the transmission range could have been extended somewhat by using much thicker wires, but the high cost of copper made that impractical.) Eventually, Edison whittled his service zone to about half a square mile, centered on the Pearl Street plant.

The building at Pearl Street was modified to support the tremendous weight of the machinery needed to generate electricity. Four heavy coal-fired boilers were placed on the ground floor to heat the water to produce pressurized steam to drive the dynamos. The basement was fitted with machines for receiving coal and removing ashes. The dynamos, or power generators, were placed on the second floor, with heavy girders and columns replacing the old flooring. The fourth floor held a bank of one thousand incandescent lights used to test the dynamos.

Laying the underground electrical wires under the streets of lower Manhattan proved to be one of the most difficult and costly jobs of all. The streets had to be torn up and trenched so that tubes containing insulated copper wires sealed with tar could be placed in the ground. It was a messy and time-consuming job, made all the more expensive by the rampant graft in New York City. As Edison recalled, "When I was laying tubes in the streets of New York, the office received notice from the Commissioner of Public Works to appear at his office at a certain hour. I went up there with a gentleman to see the Commissioner, H.O. Thompson. On arrival he said to me: 'You are putting down these tubes. The Department of Public Works requires that you should have five inspectors to look after this work, and that their salary shall be $5 per day, payable at the end of each week. Good-morning.' I went out very much crestfallen, thinking I would be delayed and harassed in the work which I was anxious to finish, and was doing night and day. We watched patiently for those inspectors to appear. The only appearance they made was to draw their pay Saturday afternoon."

Pearl Street presented a host of new problems, not least of which was finding a way to generate enough power. No dynamo then in existence could produce enough electrical energy to serve the power needs of even a half-mile square district of New York. Edison came up with what he called the "Jumbo" dynamo, named after one of P.T. Barnum's circus elephants. The Jumbo was a twenty-seven-ton behemoth that pumped out 100 kilowatts, enough to power twelve hundred lights. It was a stupendously large piece of machinery, four times the size of any available dynamo. Six Jumbos wound up being pressed into service at Pearl Street, and Edison and his men spent most of the summer of 1882 running harrowing tests on them.

"The engines and dynamos made a horrible racket," Edison remembered, "from loud and deep groans to a hideous shriek, and the place seemed to be filled with sparks and flames of all colors. It was as if the gates of the infernal regions had been suddenly opened."

Finally, the Edison system was ready for its grand debut. On September 4, 1882, at three in the afternoon, the giant dynamos at Pearl Street began to spin, sending 110 volts of direct current flashing through the underground wires to the fifty-nine customers Edison had managed to sign up by opening day.

The next day, the *New York Herald* reported: "In stores and business places throughout the lower quarter of the city there was a strange glow last night. The dim flicker of gas, often subdued and debilitated by grim and uncleanly globes, was supplanted by a steady glare, bright and mellow, which illuminated interiors and shone through windows fixed and unwavering."

The offices of financiers Drexel, Morgan & Company were among the first to be illuminated by electricity, with Edison on hand to turn on the lights in the presence of J.P. Morgan. The *New York Times* was another influential opening-day customer. The *Times* covered its own electrification in the paper: "It was not till about 7 o'clock, when it began to grow dark, that the electric light really made itself known and showed how bright and steady it is," the *Times* reported. "It was a light that a man could sit down under and write for hours without the consciousness of having any artificial light about him. . . . The light was soft, mellow, and grateful to the eye, and it seemed almost like writing by daylight to have a light without a particle of flicker and with scarcely any heat to make the head ache."

It was the dawn, not of electricity, but of the electricity business. It had come to an age scarcely prepared for electricity. It was still the era of the horse and buggy, the telegraph, and the seven-story skyscraper, of the house heated with gas or wood and illuminated with candles, kerosene lamps, and gas fixtures. Seemingly overnight, there was a new world, one in which unseen forces could do all those tasks and more. Electricity would quicken the pulse of everyday life. Edison wasn't exaggerating when he later said, "The operation of Pearl Street meant the end of one epoch in civilized life and the beginning of another."

For several months after Pearl Street opened, Edison didn't bill customers. It was better for business to give away electricity until he worked out a way to measure electricity consumption accurately, or at least to the general satisfaction of customers. Edison eventually developed an unorthodox metering device that measured current flow by chemical means. The Edison meter consisted of a jar containing two zinc plates immersed in a solution of zinc sulfate, which was connected across a shunt in the customer's circuit. When current flowed through the jar, metal was dissolved off the positive plate and deposited on the negative one. Once a month, the plates were removed from the meter by a workman, washed, and weighed on a laboratory balance. The difference in the plates' weight was a measure of the current that had been consumed. Thus, the first electricity meters weren't so much read as weighed.

Edison's meter was a quick fix, requiring a small army of men to remove and weigh the metal plates each month, and more than a few customers doubted the meter's accuracy. But the meter was a crucial component in the Edison system, letting customers be billed for exactly the amount of power they consumed, just as they had been for gas. On January 18, 1883, the first electric bill in history was sent to the Ansonia Brass & Copper Company. It was for $50.40, no doubt sparking the first electric bill complaint in history.

The early days of the Edison system were not without problems. During a thunderstorm, it was not unusual to see sparks shooting between electric chandeliers and surrounding wires. The insulation in the underground wires occasionally wore through, leaking electricity into the surrounding ground and shocking unsuspecting passers-by. Edison recalled one early mishap: "One afternoon, after our Pearl Street station started, a policeman rushed in and told us to send an electrician at once up to the corner of Ann and Nassau streets—some trouble. Another man and I went up. We found an immense crowd of men and boys there and in the adjoining streets—a perfect jam. There was a leak in one of our junction boxes, and on account of the cellars extending under the street, the topsoil had become insulated. Hence, by means of this leak powerful currents were passing through this thin layer of moist earth. When a horse went to pass over it he would get a very severe shock. When I arrived I saw coming along the street a ragman with a dilapidated old horse, and one of the boys told him to go over on the other side of the road—which was the place where the current leaked. When the ragman heard this he took that side at once. The moment the horse struck the electrified soil he stood straight up in the air, and then reared again; and the crowd yelled, the policeman yelled; and the horse started to run away. This continued until the crowd got so serious that the policeman had to clear it out; and we were notified to cut the current off. We got a gang of men, cut the current off for several junction boxes, and fixed the leak. One man who had seen it came to me the next day and wanted me to put in apparatus for him at a place where they sold horses. He said he could make a fortune with it, because he could get old nags in there and make them act like thoroughbreds."

Edison took pains to stress the safety of his system, downplaying the danger posed by the 110 volts of direct current he was sending under the streets and into people's homes. "There is no danger to life, health, or person, in the current generated by the Edison dynamo," declared an Edison circular. "The intensity of the electric current is feeble . . . in fact, the current is scarcely perceptible to the touch." But this was true only in the best of circumstances. If a poorly insulated Edison wire came in contact with, say, a metal pole, a person touching the pole would be badly shocked. Someone touching the pole while standing in a puddle of water could easily be killed. As little as 50 volts of direct current have been known to kill a human being.

Such technical matters were well beyond the grasp of the public—few could even say what electricity was. W.J. Jenks, one of Edison's first power plant managers, recalls giving two well-dressed ladies a tour of the facilities. "I invited them in, taking them first to the boiler-room, where I showed them the coal-pile, explaining that this was used to generate steam in the boiler," Jenks said. "We then went to the dynamo-room, where I pointed out the machines converting the steam-power into electricity, appearing later in the form of light in the lamps. After that they were shown the meters by which the consumption of current was measured. They appeared to be interested, and I proceeded to enter upon a comparison of coal made into gas or burned under a boiler to be converted into electricity. The ladies thanked me effusively and brought their visit to a close. As they were about to go through the door, one of them turned to me and said: 'We have enjoyed this visit very much, but there is one question we would like to ask: What is it that you make here?' "

In promoting electricity as the "modern" alternative to gas, Edison paid special attention to bringing electrical power to the homes of the wealthy. J.P. Morgan and the Vanderbilts had small electrical plants installed on their estates to provide light and power even before Pearl Street officially opened for business. A promotional circular distributed by Edison's company assured customers that replacing burned-out incandescent lamps was well within the capabilities of "an ordinary domestic." The association of electricity with affluence appealed to the social aspirations of the growing middle class in the Industrial Age. Even if you couldn't live like the Vanderbilts, you could light your home just like they did, with electricity. Not wanting to be left behind, many of New York's prominent hotels and apartment buildings quickly made the switch from gas to electrical lighting.

Even so, Pearl Street lost money for several years, mainly because the plant was so expensive to build in the first place—$300,000, counting the cost of real estate. There were significant ongoing expenses, such as the tons of coal needed to feed the plant's hungry boilers to produce the steam that drove the dynamos that produced electricity. It wasn't until Pearl Street's third year that it turned a profit.

Edison wasn't worried about absorbing a few years of losses. He was seizing the electricity market while it was still in its infancy, building demand for a commodity that, in most areas, he alone could supply. As the Edison company's 1883 annual report cheerily put it, "The Edison patents, as a matter of law, not only endow our company with a monopoly of incandescent lighting, but aside from the patents, our business has obtained such a start, one so far in advance of all competitors . . . that the business ascendancy is of itself sufficient to give us a practical monopoly."

Edison's only competition was in Europe, where the electricity market was cutting a different path. In 1882, French scientist Lucien Gaulard and his English business partner John Gibbs patented a system for distributing electricity that was fundamentally different from the Edison System in operation at Pearl Street.

The centerpiece of the Gaulard-Gibbs system was its novel power transformer, a more advanced version of the device Michael Faraday had fashioned in his laboratory half a century before. The Gaulard-Gibbs transformer could increase or decrease the voltage of the current being distributed, which gave the system an unusual degree of flexibility. In 1884, the Gaulard-Gibbs system was successfully demonstrated at an international exposition in Turin, Italy, delivering electricity to the exhibition's building.

The Gaulard-Gibbs design had another significant difference from the Edison system: it distributed alternating current, rather than direct current, AC rather than DC.

It was all still electricity, but AC and DC had different properties due to the dissimilar ways the current was generated and delivered. In the Edison DC system, the current flowed in one direction only, from the huge dynamos at Pearl Street directly to a customer's light bulb. With alternating current, the electricity flowed from the generator to the bulb, then from the bulb to the generator, flipping back and forth dozens of times per second. The alternations stem from the way an AC dynamo produces power, repeatedly cutting a magnetic field with a conducting wire so that the magnetic poles continually reverse.

AC, with its multiple changes in direction, can power a light bulb just as efficiently as DC, which delivers current only in the direction of the bulb. That's because electricity flows so fast through the wire that the light bulb filament is indifferent to a current's direction; it will illuminate either way.

The very notion that a current can so readily alternate direction is decidedly counterintuitive, which may be why Edison didn't think much of the idea when he first heard of it. "How do they make the current go the other direction?" Edison is said to have asked. A current flipping back and forth through a wire is difficult to picture, and the intensely visual Edison had no patience for it.

Edison's sole mission was to bring his DC system to cities and towns across the United States. With Pearl Street up and running, Edison began leasing his technology to start-up power companies in other locales. By leasing his technology, Edison would share in the profits of other electric companies but not be on the line financially for their success or failure. As it was, Edison could barely keep up with the demand for electrical power. By 1884, there were eighteen central stations on the Edison system, producing DC power for cities including Chicago, Boston, Philadelphia, and New Orleans.

In the span of a few years, Edison had built an electrical empire out of thin air; now the angels in the wire were dancing to his tune. A mighty current had been unleashed, and it seemed as though it would flow on forever.

6

TESLA

On a luminous day in the summer of 1884, a stranger strode through the doors of 65 Fifth Avenue and introduced himself to Edison as a new employee. The meeting was unremarkable; the stranger was anything but.

Edison sized up the newcomer—tall, dark-haired, thin as a rail, with raccoon-like circles under his eyes. The visitor's eyes were blue-gray, like Edison's, but there was something unusual about his gaze, a far-off look that one rarely saw among the hard-charging practical men that usually came to Edison's door. The stranger handed Edison a letter of introduction, identifying himself as Nikola Tesla, a Serbian electrician from the Continental Edison company in France. For Tesla, merely standing in the same room with Edison was one of the singular thrills of his career.

"The meeting with Edison was a memorable event in my life," Tesla later recounted. "I was amazed at this wonderful man who without early advantages and scientific training had accomplished so much. I had studied a dozen languages, delved in literature and art, and spent my best years in libraries . . . and felt that most of my life had been squandered."

Negative and positive. Their natures were as opposite as the poles of the dynamos spinning away in the basement of Pearl Street. Tesla was a dreamy twenty-eight-year-old immigrant trying to find his way in the New World. Edison was only four years older, but the age difference seemed much greater. Edison's hair had already started turning white; his eyebrows were bushy, protruding from his forehead like the bow of a ship. His employees already referred to him as the Old Man.

Tesla loved mathematics and abstract thinking; Edison hated math and preferred to work on problems that could be easily visualized. Edison had an orderly and supremely rational mind, capable of juggling dozens of inventions while simultaneously running a large business. Tesla's mind was more like lightning; his insights were brilliant, unpredictable, and not always on target. Edison had cultivated a folksy persona given to pithy sayings that endeared him to the public. Tesla was a bundle of raw nerves and runaway phobias.

Opposites attract, particularly in electricity. Tesla would wind up working for Edison for less than a year, but the two men would be linked forever by fate and electricity, by AC and DC.

Nikola Tesla was born at precisely midnight, so the story goes, between July 9 and 10, 1856, in the village of Smiljan, Croatia. It was as though time itself bent to Tesla, his arrival neatly straddling the alternating cycles of the day. Tesla's father was the pastor of the local Serbian Orthodox Church, and from birth, young Nikola was intended for the clergy. Tesla later said that the prospect of becoming a minister "hung like a dark cloud on my mind."

Nikola went his own peculiar way. From an early age, he was visited by strange apparitions. Unexplained and often unwanted images would suddenly appear to Tesla, so lifelike they blocked his vision of real objects. Tesla would later describe the condition: "In my boyhood I suffered from a peculiar affliction due to the appearance of images, often accompanied by strong flashes of light, which marred the sight of real objects and interfered with my thought and action. They were pictures of things and scenes which I had really seen, never of those I imagined. When a word was spoken to me the image of the object it designated would present itself vividly to my vision and sometimes I was quite unable to distinguish whether what I saw was tangible or not. This caused me great discomfort and anxiety."

To free himself of the troublesome images, Tesla tried replacing them with other mental pictures. But the relief was only temporary; Tesla found it exhausting coming up with new scenes to fill his mental reel of pictures. He would later complain of lightning-like prismatic images whenever he closed his eyes.

Tesla also suffered from what today would be diagnosed as obsessive-compulsive disorder. He had a lifelong germ phobia that led him to develop complex rituals to alleviate his fears of contamination. He went to great lengths to avoid shaking hands, placing his hands behind his back when anyone approached. If a visitor caught him off guard and forced him to shake his hand, Tesla would dismiss the guest, rush to a washroom, and scour his hands. Workmen eating their lunch with dirty hands nauseated him. When dining at a restaurant, he required that a fresh tablecloth be put down and that other patrons not use his table. He insisted that the utensils be sterilized, and even then, carefully wiped them down with as many as two dozen napkins. If a fly alighted while Tesla was eating, he would scurry off to another table, and the entire ritual had to be repeated. Tesla was also repulsed by the sight of objects with smooth surfaces, particularly pearls. A woman wearing a string of pearls in the same restaurant was enough to send him out the door.

"Tesla was not oblivious of his idiosyncrasies," wrote John O'Neill, one of Tesla's associates who later wrote a book about him. "He was quite aware of them and of the friction which they caused in his daily life. They were an essential part of him, however, and he could no more have dispensed with them than he could his right arm. They were probably one of the consequences of his solitary mode of life or, possibly, a contributing cause of it."

Tesla's mind was unruly, mercurial, and utterly original. As a boy, Tesla followed an impulse that would later come to dominate him—a desire to harness the power of nature and put it to work. One of his earliest experiments was to attach several June bugs to a thin wooden spindle. The motion of the bugs' legs was transmitted to a large disk, making it rotate. It was the first Tesla motor.

Early on, Tesla was fascinated by electricity. Tesla stroked his cat's back one day and was amazed to see its fur emit a shower of sparks. "My father remarked this is nothing but electricity, the same thing you see on the trees in a storm," Tesla remembered. "My mother seemed alarmed. 'Stop playing with the cat,' she said, 'he might start a fire.' I was thinking abstractly. Is nature a cat? If so, who strokes its back? It can only be God, I concluded. I cannot exaggerate the effect of this marvelous sight on my childish imagination. Day after day I asked myself what is electricity and found no answer."

To better understand electricity and how it could be generated, Tesla made a careful study of the mechanical models of electrical turbines that his school had on display. Tesla constructed makeshift water turbines of his own and took great pleasure in watching them spin in local creeks.

"My uncle had no use for this kind of pastime and more than once rebuked me," Tesla recalled. "I was fascinated by a description of Niagara Falls I had perused, and pictured in my imagination a big wheel run by the falls. I told my uncle that I would go to America and carry out this scheme."

Tesla's uncle laughed. Thirty years later, the boy would make good on his word.

Like many born inventors, Tesla received low marks in school. His favorite subject was math, and he was so adept at mental arithmetic that some teachers suspected him of cheating when he calculated complicated mathematical problems without picking up a pencil. Languages also came easily; to his native Serbo-Croat, Tesla added German, Greek, Italian, French, and English.

Tesla entered college at age fifteen at Karlovac in Croatia. He completed the four-year program in three years, and in 1875 enrolled at the Polytechnic Institute in Gratz, Austria. One of Tesla's favorite teachers at the Institute was professor Jacob Poeschl, a methodical and literal-minded German who was chair of the physics department. Poeschl could match Tesla for peculiarity; it was said that Poeschl wore the same coat for twenty years. But the professor had a relentless attention to detail that Tesla grew to admire. "I never saw him miss a word or gesture, and his demonstrations and experiments always went off with clock-like precision," Tesla recalled.

Poeschl obtained a small DC motor from Paris, and he used it to demonstrate various effects of direct current to his students. One day, the motor malfunctioned—the copper wire brushes that made and broke contact with the rotating mechanical commutator began to throw off a shower of sparks.

The commutator and brushes were and are an essential element of DC motors. They form a switching mechanism that reverses the current twice during each rotation of the rotor so that the opposing north and south magnetic fields keep the rotor spinning continuously, north to south, negative to positive. When the timing of the switching is off, the brushes spark, and the motor loses power or stops altogether. The heavier the load on the motor, the worse the problems get.

Tesla thought the whole commutator setup was inefficient—you'd never see a design like that in nature, he thought. During class, he proposed building a motor without any commutator. Professor Poeschl listened attentively to his bright young pupil, and after Tesla was done, the professor loudly declared that such a machine was absolutely impossible. Poeschl then devoted an entire class lecture to enumerating the numerous ways in which Tesla's proposed motor violated fundamental laws of physics. "Mr. Tesla may accomplish great things, but he certainly will never do this," Poeschl declared. "It is a perpetual motion scheme, an impossible idea."

At first, Tesla was chastened by his professor's rebuke. But soon he began to wonder whether the idea was impossible after all. For not the last time in his life, Tesla followed his intuition.

"I could not demonstrate my belief at the time," Tesla later recalled. "But it came to me through what I might call instinct, for lack of a better name. We undoubtedly have in our brains some finer fibers which enable us to perceive truths which we could not attain through logical deduction. . . . I undertook the task with all the fire and boundless confidence of youth. To my mind it was simply a test of will power. I knew nothing of the technical difficulties."

Tesla applied his formidable powers of abstraction to the task. Motor designs danced in his head. Devices were mentally assembled, taken apart, and put back together.

"I started by first picturing in my mind a direct-current machine, running it and following the changing flow of the currents in the armature," Tesla said. "Then I would imagine an alternator and investigate the progresses taking place in a similar manner. Next I would visualize systems comprising motors and generators and operate them in various ways."

The images were palpable; Tesla could build fully formed worlds in his mind. Tesla's remaining time at the Polytechnic Institute was spent obsessing over electric motors. "I almost came to the conclusion that the problem was insolvable," he said.

In 1880, Tesla moved to Prague and landed a job as chief electrician of the city's new telephone company. While strolling through City Park one late afternoon, Tesla was transfixed by the sight of a dramatic, blood red sunset. At that moment, the sun seemed to him to be a swirling ball of energy, a gigantic rotating magnetic field. A passage from Goethe's *Faust*, which Tesla knew by heart, sprang to mind:

The glow retreats, done is the day of toil
It yonder hastes, new fields of life exploring;
Ah, that no wing can lift me from the soil
Upon its track to follow, follow soaring!

It was an epiphany. Tesla said, "The idea came like a flash of lightning, and in an instant the truth was revealed." He immediately picked up a stick and began drawing diagrams in the sand. The drawings would form the basis of a breakthrough patent Tesla received in May 1888.

What Tesla had come up with was the induction motor, a new and vastly more efficient motor design that did away with the commutator altogether. Instead of copper brushes constantly rubbing against metal to change the magnetic poles in the rotor, Tesla's motor was spun by rotating the magnetic field itself—an idea suggested by the swirling magnetic field Tesla imagined in the Prague sunset. The magnetic field could be induced to rotate if two coils set at right angles were supplied with an alternating current. The induction motor was almost magical; it operated without any moving electrical contacts, driven instead by an invisible magnetic field. It was the sort of elegant simplicity found in nature, Tesla's constant source for inspiration.

Tesla's AC induction motor represented a more direct application of electrical energy to spin a rotor. There were no brushes to wear out or spark, no external commutator to slow things up. By rapidly changing the rotating magnetic field, the Tesla motor could be spun in one direction, stopped on a dime, and rotated the other way just as quickly. It was a design of such grace that many scientists and electricians would later wonder why they hadn't thought of it themselves.

Tesla constructed a crude version of his induction motor in 1883, letting him see for the first time a motor powered by an alternating current without the use of a commutator. Tesla was more convinced than ever that he was on the right track, but he was unable to raise enough money to build a proper prototype of his motor.

Tesla took a job with Continental Edison near Paris, a French company making dynamos, lamps, and motors for European markets under Edison's patents. He came to the attention of Charles Batchelor, a longtime Edison assistant and manager of the plant. Batchelor encouraged Tesla to go to America and work for Edison directly, and gave the young electrician a letter of introduction. In the summer of 1884, Tesla sailed to New York City and landed with virtually no possessions to declare. Everything he had of value was stored in his head.

Tesla went to see Edison at the inventor's headquarters on Fifth Avenue. One account has Tesla producing Batchelor's letter of introduction to Edison, which supposedly read: "I know two great men and you are one of them; the other is this young man." More likely, the letter simply vouched for Tesla's technical expertise without going so far as to compare him to the world-famous inventor of the light bulb and phonograph.

Edison was not a man easily impressed by the opinion of others, anyway. The measure of a man was his ability to get the job done.

Tesla's first assignment was to fix the lighting plant for the S.S. *Oregon*, the fastest passenger steamer of its day. Both of the ship's DC dynamos were disabled, delaying the departure of the craft and creating unfavorable publicity for the Edison system. Tesla managed to repair both dynamos in one long evening's work. When Tesla told Edison he had just come from the *Oregon* and had repaired both machines, the inventor looked Tesla square in the eye and walked away without saying a word. But Tesla later heard Edison remark, "This is a damn good man."

"Within a few weeks I had won Edison's confidence," Tesla recalled. He was given unusual freedom to design DC dynamos and motors and put in grueling hours. For nearly a year, Tesla's regular workday stretched from ten-thirty one morning until five o'clock the next. At one point, Edison took Tesla aside and said in his reedy voice: "I have had many hardworking assistants but you take the cake." Coming from the Old Man, it was high praise indeed.

Tesla and Edison were hardly equals; the two men inhabited vastly different worlds. Edison was the renowned Wizard of Menlo Park, master of a sprawling electrical empire, a man of social standing in New York. Tesla was an unknown electrician who spoke with a strange accent, and whose sole invention existed entirely in his head.

The two men's work methods were also markedly dissimilar. Tesla preferred to go for months and even years with an idea slowly taking shape in his mind. By the time Tesla got around to making a sketch on paper, the invention had been fully worked out in his head. "Without having drawn a sketch I can give the measurements of all parts to workmen, and when completed all of these parts will fit, just as certainly as though I had made the actual drawings," Tesla said.

Edison's methods couldn't have been more different. "If Edison had a needle to find in a haystack, he would proceed at once with the diligence of the bee to examine straw after straw until he found the object of his search," Tesla later recalled, with some annoyance. "I was a sorry witness of such doings, knowing that a little theory and calculation would have saved him ninety percent of his labor."

In the less than ten months Tesla worked for Edison, there were only rare social encounters between the two men, such as the time Edison bet Edward Johnson, president of the Edison Illuminating Company, he could guess Tesla's weight.

"Someone suggested guessing weights and I was induced to step on a scale," Tesla recalled. "Edison felt me all over and said: 'Tesla weighs 142 lbs. to an ounce," and he guessed it exactly. Stripped I weighed 142 lbs. and that is still my weight. I whispered to Mr. Johnson: 'How is it possible that Edison could guess my weight so closely?' 'Well,' he said, lowering his voice. 'I will tell you, confidentially, but you must not say anything. He was employed for a long time in a Chicago slaughter-house where he weighed thousands of hogs every day.'"

All the while, Tesla ached to tell Edison about his induction motor. Tesla knew that Edison didn't think much of alternating current. It was all nonsense, Edison said, an unproven and unreliable system favored by Europeans who didn't know the first thing about electricity. But Tesla held out hope that Edison would see the beauty of the induction motor's simple design and overcome his prejudice against alternatives to his DC system. Tesla finally worked up the nerve to approach Edison about the induction motor when the two men were at Coney Island in late 1884.

"The moment that I was waiting for was propitious," Tesla recalled. "I was just about to speak when a horrible looking tramp took hold of Edison and drew him away, preventing me from carrying out my intention."

The story is just odd enough to be true. Or perhaps Tesla never worked up the courage to approach Edison. Or if he did, Edison rejected the idea out of hand. After all, Tesla's motor wasn't an improvement to the DC system; in many ways it was a repudiation of it. The two electrical systems were completely incompatible; a motor could be built to run on AC or DC, but not both.

In any event, Tesla quit the Edison Works in spring 1885, ostensibly over a $50,000 bonus Tesla had been promised that never came through. But it was more that Tesla simply didn't fit. Tesla and

Edison were far too different to strike up a working partnership. Both men were geniuses in their own way, but Edison's was 99 percent perspiration, Tesla's 99 percent inspiration. The two men would rarely cross paths again, but their inventions would soon clash openly in the marketplace.

Tesla initially floundered after quitting Edison. He took a job as a ditch digger for a while, and would later recall this period with considerable embarrassment. Gradually, Tesla found his footing and began to shop his induction motor idea to potential investors. That's how he met George Westinghouse.

"My first impression of Westinghouse was that of a man with tremendous potential energy of which only part had taken kinetic form," Tesla recalled. "A powerful frame, well proportioned, with every joint in working order, an eye as clear as crystal, a quick and springy step—he presented a rare example of health and strength. Like a lion in a forest, he breathed deep and with delight the smoky air of his factories."

George Westinghouse was a Pittsburgh-based inventor and industrialist famous for devising the railroad air brake, a safety device that saved countless lives. Westinghouse was a bear of a man, a large-framed figure with a walrus mustache, genial manner, and the sort of stoutly reliable face you'd see on a box of cough drops.

Born in 1846, a year before Edison, Westinghouse was raised in an atmosphere of invention. His father had a bustling farm machinery shop and was awarded seven patents for threshing and sawing machines. Young George received poor grades in school—another underachieving future inventor—and preferred to tinker in his father's workshop. He learned to read blueprints at an early age and began to sketch his own designs.

At seventeen, Westinghouse ran off to join the Union Army in the Civil War. He eventually transferred to the Navy and worked as an engineer on two steam-powered battleships. After the war, Westinghouse turned to invention full time and was awarded his first patent in 1865 for a rotary steam engine. He had just turned nineteen.

Three years later, Westinghouse came up with what would be his most famous invention, the railroad air brake. It was built on the idea of applying braking to all wheels of railroad cars by means of compressed air driven by a steam pump. Westinghouse's air brake system transformed the railroad industry, significantly reducing accidents. Before the Westinghouse brake, it took nearly a mile to stop a fully loaded passenger train going only ten miles per hour. With the Westinghouse brake, a train traveling thirty miles per hour could be halted in just five hundred feet. Trains could take on larger and heavier loads because of their improved stopping distance, greatly expanding the reach of the railroad.

Westinghouse eventually became a world-class inventor in his own right, credited with nearly four hundred patents. But Westinghouse didn't give himself over completely to invention the way Edison did. Westinghouse's avuncular nature made him a natural dealmaker. He enjoyed directing the work of other men, adapting existing ideas, combining companies, buying up patents, assembling conglomerates. To Westinghouse, building a business was an act of invention, every bit as much as coming up with the air brake.

In the early 1880s, Westinghouse began to turn his attention to electricity. Westinghouse had been one of thousands of spectators at Edison's Menlo Park lab when the Wizard first demonstrated his incandescent lamp. Electricity had always been technically interesting to Westinghouse; with the success of the Edison system, it now seemed potentially profitable. In December 1885, Westinghouse joined with his brother and a handful of other backers to form the Westinghouse Electric Company, with capital stock of $1 million. The main assets of the company were twenty-seven patents relating to electricity that Westinghouse had bought up.

Most of the patents purchased by Westinghouse were for direct current lighting and power systems. The designs were similar to the Edison system but just different enough to avoid obvious patent infringement. Westinghouse installed a small-scale isolated DC incandescent lighting plant for the Windsor Hotel in New York City in 1886, and soon after lit the Monongahela Hotel in Pittsburgh, the city where his company was headquartered. Later the same year, the first Westinghouse central station opened at Trenton, New Jersey, generating DC power from six Siemens dynamos. Westinghouse followed up with additional DC generating plants in Plainfield, New Jersey, and Schenectady, New York.

The DC market was tough to crack, though. The Edison companies dominated the industry; customers knew and trusted the Edison name. Edison controlled all the best patents on DC lamps, dynamos, and motors, and his company became increasingly aggressive about filing lawsuits against suspected patent infringers.

With Edison's near-monopoly of the DC market, Westinghouse turned his sights to the new technology emerging in Europe: alternating current. AC transmission was largely unproven, but it had some interesting qualities that allowed it to outperform DC in certain situations.

One of DC's biggest shortcomings was that it couldn't be transmitted much more than a mile from the central station without significantly losing power. Edison's Pearl Street station barely served half a square mile of New York; dozens of stations would have to be built to serve the entire city, and real estate was expensive in New York. Sparsely populated areas might never be electrified, since no company was going to build a DC power plant to serve a handful of people.

Alternating current, on the other hand, could be made to travel farther, thanks in part to the transformer. With the transformer, alternating current could be easily increased or "stepped up" to a higher voltage, which could travel through wire more easily. Consequently, high-voltage AC could be transmitted longer distances along thinner, cheaper copper wire, with the voltage then "stepped down" for use in homes and offices.

There was no practical way to increase and decrease DC voltages. Direct current was best produced and transmitted at a low, constant current, 110–220 volts, and thus didn't have AC's built-in flexibility.

Westinghouse was intrigued by AC's potential but was unsure whether it was reliable or cheap enough to rival DC. Articles in the electrical trade journals were regularly hostile to AC, dismissing it as an unnecessary and unworkable alternative, a laboratory trick best kept in the laboratory. Critics contended that in stepping up voltages to several thousand volts to transmit the power, much of the energy would be lost in the form of heat. A full-blown AC system would be nothing more than a gigantic heater and a complete disaster for its investors.

Westinghouse called on Frank Pope, one of his most trusted electrical experts, to investigate the alternating current system. Pope was an AC skeptic, but studying the system more closely changed his mind.

"My own impression at first sight was, like that of every one else, an unfavorable one," Pope later recalled. "The knowledge which I had gathered in the ordinary course of my professional experience led me to expect that the loss of energy in conversion would be so great as to render the scheme commercially unprofitable, and that this lost energy, appearing in the form of heat, would quickly destroy the apparatus or at least render it useless. It was not until I had gone through the published researches . . . that I found reason to change my opinion. I followed up on the matter . . . and was convinced of its novelty and industrial value."

Persuaded that AC was worth a gamble, Westinghouse went out and bought the best AC patents he could find, the Gaulard-Gibbs system from Europe. It wasn't a full alternating current system, but Gaulard-Gibbs had an essential piece: the transformer that stepped up and stepped down the line voltage, the key to AC's cheap long-distance transmission. A version of the transformer was brought to Pittsburgh, and Westinghouse and his team of engineers set out to improve the design.

There were no rules governing how to build an alternating current system; the Westinghouse team was making them up as they went along. Some of the technical decisions they made then remain with us more than a century later, such as having the current alternate at 60 cycles per second, still the standard in North America.

Westinghouse's chief engineer, William Stanley, designed a complete alternating current system in 1886 to bring electricity to a handful of stores and offices in the tiny town of Great Barrington, Massachusetts. It was the first working AC transformer installation in the country. Twelve transformers, newly designed by Stanley, stepped down 3,000 volts of alternating current in the transmission line to 500 volts, which illuminated four hundred incandescent lamps in the sleepy Berkshire town. Eight months later, Westinghouse opened his first commercial AC plant in Buffalo, New York, and soon had orders to build more than two dozen additional AC central stations. By the end of 1886, Westinghouse Electric employed three thousand people, still considerably fewer than Edison's global electrical empire. But Westinghouse was becoming a significant rival, and a growing threat to Edison.

The market for alternating current, however, still faced a roadblock. Westinghouse was missing a crucial piece of a complete electrical system: a reliable motor that would run on AC. Nearly all the commercial motors made at the time ran on Edison's DC system; it would have been foolish for manufacturers to make anything else. The few alternating current motors available were markedly inferior; they couldn't start by themselves and were prone to vibrate wildly once they were running.

Nikola Tesla, induction motor in hand, came along at just the right time for Westinghouse. Tesla had been shopping his induction motor for two years after leaving Edison, with no success. On May 1, 1888, he was awarded a series of patents, among them U.S. patent number 381,968 for an "Electro magnetic motor," and patent number 382,280 for "Electrical Transmission of Power." The latter patent detailed how alternating current could be used to drive the motor, what would become known as the "Tesla polyphase system." It was called *polyphase* because it employed multiple currents, each out of phase, or step, with the others. It was a bit like adding multiple pedals to a bicycle—when one pedal reached the bottom of its stroke, another pedal reached the top and began to push down, producing a steady flow of power. By using out-of-phase alternating current, there was always one cycle nearing its peak.

Around the same time, Tesla accepted a last-minute invitation to give a lecture on his work before the American Institute of Electrical Engineers at Columbia University. The lecture, titled "A New System of Alternating Current Motors and Transformers," turned out to be a sensation. Tesla demonstrated two small induction motors and many of the academics in attendance reacted with astonishment and even a touch of chagrin. The induction motor was simplicity itself, rotating without any moving electrical contacts. It made DC motors look clunky by comparison. The lecture and demonstration established Tesla's name in the scientific community practically overnight.

George Westinghouse contacted Tesla just days after his breakthrough lecture. Westinghouse knew that if the AC induction motor was everything Tesla said it was, it could be the reliable motor he had been looking for, the missing piece in his commercial AC system. After some negotiation, Westinghouse bought the rights to Tesla's patents for $70,000 plus a royalty of $2.50 per horsepower for each Tesla motor. Once the deal was signed, Tesla moved to Pittsburgh and worked beside Westinghouse for nearly a year, adapting the Tesla motor to the Westinghouse system.

During his time with Westinghouse, Tesla grew to admire the inventor-industrialist. Westinghouse might not have been the creative equal of Tesla's previous boss, but he was both fair-minded and fiercely competitive, a rare combination.

"Always smiling, affable and polite, he stood in marked contrast to the rough and ready men I have met," Tesla said. "Not one word which would have been objectionable, not a gesture which might have offended—one could imagine him as moving in the atmosphere of a court, so perfect was his bearing in manner and speech. And yet no fiercer adversary than Westinghouse could have been found when he was aroused. An athlete in ordinary life, he was transformed into a giant when confronted with difficulties which seemed insurmountable. He enjoyed the struggle and never lost confidence. When others would have given up in despair he triumphed."

Edison viewed the Tesla-Westinghouse collaboration with mounting suspicion. The inventor had nothing against Tesla for selling his induction motor to Westinghouse. The AC motor, along with the entire alternating current system, Edison believed, was doomed to failure. "Tesla is the poet of science," Edison declared, a maker of "magnificent but utterly impractical" inventions. Edison retained a measure of fondness for Tesla, whom he admired for his creativity and hard work, even if many of Tesla's ideas seemed not fully grounded in reality.

Westinghouse was another matter; Edison quickly grew to hate him. It wasn't so much Westinghouse the man Edison detested. The genial industrialist was difficult to dislike personally; in an age of ruthless robber barons, Westinghouse was practically a saint. Rather, Edison hated what Westinghouse had come to represent to him: the intrusion of moneymen into science, the amassing of electrical empires by men who knew nothing about science or technology and didn't care to, as long as the profits kept pouring in. In the electricity business, the suits were increasingly calling the shots, not the men in grimy overalls, and this rankled Edison deeply. Westinghouse became the symbol for Edison of all that was wrong about the world of business.

Edison groused, "Just as certain as death Westinghouse will kill a customer within six months after he puts in a system of any size. He has got a new thing and it will require a great deal of experimenting to get it working practically. It will never be free from danger."

Edison went public with his opinion in a pamphlet published in 1886. "A WARNING FROM THE EDISON ELECTRIC LIGHT COMPANY," the cover declared in blood-red lettering. The pamphlet was ostensibly a warning to would-be patent infringers, promising swift legal action. But its dual purpose was to stir up fear over the hidden dangers of the newfangled alternating current.

The use of AC meant "greatly enhanced risks to life and property," the pamphlet declared, cautioning that the cost of such damages would have to be borne by those who purchased a Westinghouse AC power plant. The brochure included several graphic newspaper accounts of accidental deaths at the hands of alternating current. In one, an electrical lineman was found grotesquely hanging by his neck in a nest of electrical wires sixty feet off the ground after having been dealt a fatal shock by a Westinghouse line. In another, a theater manager was struck dead on the stage in the middle of a Saturday matinee when he received a fatal shock of AC from a poorly insulated wire.

No such horrors occurred with DC systems, the Edison pamphlet assured consumers. In contrast to the "deadly" AC system, "we have the glorious record of the Edison low tension system, from which there has never been a single instance of loss of life from the current employed." The brochure confidently predicted that the AC system "is not destined to assume any permanent position. It would be legislated out of existence in a very brief period even if it did not previously die a natural death."

The shrillness of the attack revealed more than Edison intended—the challenge from AC had him spooked. Since the early days of Pearl Street, Edison had enjoyed a comfortable monopoly on electrical generation and distribution, and had seen small rivals come and go. But AC was different. It didn't seek to improve upon Edison's DC system; it aimed to usurp it. The Old Man never walked away from a fight, and this was shaping up to be a bare-knuckle brawl.

7

THE ANIMAL EXPERIMENTS

The electrical industry was still in its infancy, but already, ambitious young men were promoting themselves as experts in the field. One such self-appointed authority in New York City had smart-looking business cards printed up to announce his credentials to the world. The card read:

> Harold P. Brown
> Electrical Engineer
> 45 & 47 Wall Street
> New York

Harold Brown was attracted to electricity like a moth to a flame. As a young man, he was caught up in the excitement following Edison's incandescent lamp breakthrough and threw himself into the electricity business, even though he had no prior experience in the field. He landed a job with the Western Electric Company in Chicago, which sold devices powered by the Edison DC system. At Western Electric, Brown was put in charge of promoting one of Edison's less-celebrated inventions, the electric pen, an early stenciling device. Brown saw himself as more than a mere salesman, however. In December 1879, he wrote to Edison, claiming to "have personally sold most of the [electric pens] that have been disposed of west of New York," and to be "therefore better posted on the subject of the duplicating business than anyone else." There is no record of Edison replying to Brown; Edison received hundreds of letters from ambitious young men seeking to get in on his action.

After two years at Western, Brown joined the Brush Electric Company, the company that had designed the arc lighting system for Wanamaker's department store in Philadelphia. Brown would later characterize his role at Brush as an "electrical expert" but it appears much of his time was spent as a salesman, hawking arc lighting systems to businesses around Chicago.

Brown then decided to strike out on his own as an inventor, emulating his idol, Thomas Edison. Brown came up with several safety improvements to arc lighting systems and tried to have them patented. But after four years of fruitless patent battles, Brown decided he wasn't cut out for the invention game. Instead, he bestowed a new title on himself, one that was starting to make the rounds in the industry: *electrical engineer.*

Like many who claimed the title, Brown had only a rudimentary knowledge of electricity and no formal training as an engineer; he had only a high school education. But it hardly mattered. Brown's self-interest outstripped his judgment and his ambition outran them both. It didn't take much to be an expert in the field; the public was, for the most part, utterly ignorant of electricity and how it worked. Brown simply proclaimed himself an electrical engineer and set up shop in the heart of New York's financial district. One of his specialties was modifying arc lamp dynamos so they'd be somewhat less likely to give a fatal jolt to an unsuspecting operator.

The business of electricity was booming. The success of Edison's Pearl Street plant had encouraged others to enter the market, even though many had no prior experience with electricity. New electrical lines were being strung in New York every week, faster than they could safely be installed. Some installations were crude patch jobs, with poorly insulated wires snaking around existing telegraph and telephone lines, the mad tangle literally leaking electricity. The New York newspapers began to feature a recurring story: the death by electricity of an unsuspecting victim. The articles were accompanied by sensational headlines such as "THE WIRE'S FATAL GRASP" and "AGAIN A CORPSE IN THE WIRES."

Harold Brown read the articles and saw not danger but opportunity. In June 1888, he wrote a strongly worded letter to the editor of the *New York Post,* blaming a string of recent electrical deaths on the use of alternating current. Brown's entire career had been built on selling and servicing direct current systems, and he minced no words in describing the rival AC standard to the reading public.

"The alternating current can be described by no adjective less forceful than damnable," Brown fulminated. "The only excuse for the use of the fatal alternating current is that it saves the company operating it from spending a larger sum of money for the heavier copper wires which are required by the safe incandescent systems. That is, the public must submit to *constant danger from sudden death,* in order that a corporation may pay a *little larger dividend.*"

Placing the AC wires underground would only make matters worse, Brown contended, "as dangerous as a burning candle in a [gun] powder factory." Brown's letter concluded with a list of pointedly self-serving recommendations, among them that arc light systems be required to carry a host of new safety features, ones that Brown just happened to be in the business of providing.

The *Post* letter immediately catapulted the anonymous electrician into the growing public debate over safety. Brown was invited to appear before the New York Board of Electrical Control, a newly formed body empowered to regulate the city's unruly electrical industry. Perhaps sensing that the board could not be won over with an emotional appeal, Brown submitted a more measured critique of alternating current, even stating at one point that no electrical system was inherently safer than another. Nevertheless, Brown blamed high-voltage AC for the recent spate of accidental deaths in the city, and advanced an audaciously self-serving recommendation: that alternating current in New York be limited to 300 volts.

This proposal struck at the heart of AC's rapid rise as a competitor to the Edison system. The chief advantage of alternating current was that it could be transmitted greater distances, thanks to the high pressure of 1,000 volts or more. Limiting AC to a maximum of 300 volts took away its chief economy. Such a low voltage would require three times more copper wire to carry, effectively pricing AC out of the market.

Under the guise of protecting the public from the danger of AC, Harold Brown pushed for regulations that would protect his own DC-based line of work. The more stringent arc lighting standards he proposed would bring him a flood of new work, and a 300-volt limit on alternating current would effectively cripple the chief rival to Brown's primary source of business.

Brown's proposals were brought up at a June meeting of the Board of Electrical Control, and invitations were extended to the various electric-lighting companies to debate the matter the following month. The follow-up meeting turned out to be a rancorous affair. Men from the DC power companies and arc light concerns heaped praise on Brown's proposals, loudly declaring that public safety was at stake. The AC forces made impassioned statements in support of their standard and bitterly denounced Brown as a stooge for DC interests.

The thin-skinned Brown took the criticism personally, complaining that the meeting made him "the subject of the most violent personal abuse" and that his opponents had "done all they could do to publicly blast my reputation and stamp me as an ignorant imposter in electrical engineering." Brown saw the attacks as an assault on his carefully cultivated image as an electrical "expert," which only made him harden his position. He would never again concede that direct current could be just as dangerous as alternating current. From now on, Brown was out to prove that AC was a "damnable death current" while DC was "completely harmless."

"There remained but one thing for me to do to clear myself," Brown recounted. "I must show from their own current and its effect upon life as compared with continuous currents that my statements are true. Words are of no avail against such accusations as theirs."

Brown proposed a series of experiments to compare the relative dangers of AC and DC in a way that could be easily understood by the public. To give his demonstrations credibility as a scientific endeavor, Brown impulsively called upon Edison at the inventor's new laboratory in West Orange, New Jersey, and asked for a loan of electrical instruments.

"To my surprise, Mr. Edison at once invited me to make the experiments at his private laboratory, and placed all necessary apparatus at my disposal," Brown said.

The two men hadn't met before, although they had been working on parallel tracks for some time, each man promoting the "safe" direct current over the "deadly" alternating variety. Once introduced, Edison and Brown quickly became not so much friends as accomplices. In Edison, Brown gained a powerful and widely respected benefactor in his fight against AC. In Brown, Edison found a man willing to do almost anything to advance the cause.

Brown would always maintain that Edison never hired him, stating under oath in a later court case, "I have never been employed by any Edison electric light interest." Brown may not have been officially on the Edison payroll, but it's clear he received significant support from the inventor, in both money and access to the lab's equipment and expertise. (An enterprising *New York Sun* reporter would later get his hands on a cache of papers stolen from Brown's Wall Street office that documented Brown's close business relationship with Edison.) For Brown, the association with Edison provided him with something he could never hope to acquire on his own: respect.

That Edison agreed to team up with the unscrupulous Brown is a measure of how worried the inventor had become over AC's steady inroads into his electrical distribution empire. Edison's DC system still had more power plants, but the Westinghouse AC system was adding new plants at a faster rate. Edison was also being squeezed by a French syndicate that was cornering the market on copper, sending the already high price of copper soaring. The fact that high-voltage AC could be transmitted using thinner, cheaper wire made it even more attractive when copper prices spiked. Edison had fought off scores of competitors in his day, but the AC forces were proving to be formidable opponents.

Edison had no scientific evidence that AC was inherently more dangerous than DC, despite his company's claims to the contrary. The anecdotal evidence from dozens of accidental electrical deaths suggested that either current could kill under the right circumstances. Higher voltages certainly posed a greater threat to life and limb, but DC-powered arc light systems had been using 3,000 volts for years without a word of protest from Edison. The Westinghouse AC system used at most 2,000 volts, and that was confined to street lines. The alternating current going into homes and offices was stepped down to as little as 50 volts, while Edison's DC system ran a 110-volt DC line into the home. Edison's dire warnings about the dangers of AC were built more on fear than facts. Harold Brown's experiments might demonstrate something more tangible to support his claims.

At the same time, Edison must have sensed something in Brown that gave him pause; from the start, the inventor kept his relationship with the ambitious electrician at arm's length. Edison assigned his chief electrician, Arthur E. Kennelly, the task of assisting Brown in his experiments, which would be performed at the Edison laboratory. Edison's lab had all the instruments an electrical experimenter could ever want; all Brown needed now were some subjects.

In early July 1888, the word went out on the streets of Orange that the Edison lab would pay 25 cents for every stray dog delivered to its door. Neighborhood boys led the roundup, and the lab soon had more than enough subjects for Brown's experiments. (Brown briefly considered using cats as subjects but decided against it because, as he explained, "The cat is very apt to wiggle around when you attempt to apply the electrode, and they also have claws.")

The experiments at the Edison lab began at ten o'clock on the evening of July 10, under the soft glow of the incandescent lamps that Edison had invented nearly a decade before. Brown had set up a portable dynamo capable of generating 1,500 volts attached to a pair of wires that would be attached to each dog's legs. Brown detailed the proceedings in a lab notebook, setting down in dispassionate prose the torture and execution of living creatures in the name of science.

First Experiment

> Dog No 1. Old black and tan bitch; low vitality; weight not taken (about 10 lbs.). Resistance from right front leg to left hind leg, 7,500 ohms. Connection made through roll of wet cotton waste, held in place by wrappings of bare copper wire; continuous [direct] current used. Electromotive force at time of closing circuit 800 volts; time of contact through dog 2 seconds.

When Brown closed the circuit, 800 volts of direct current surged into the black and tan dog. The animal let out a piercing howl and made a violent effort to escape, proof to Brown "that it had control of its muscles and that nerve functions were not destroyed." After two seconds, the circuit was interrupted, and the dog howled even louder. It continued to wail and rush around in pain for two and half minutes before it finally dropped on its side in a heap. Twenty-one minutes after the dog received the shock of DC, its heart stopped beating. It was time for a new dog.

Second Experiment

> Dog No. 2. Large half-bred St. Bernard puppy; strong and in good condition. Weight not taken (about 20 lbs.). Resistance from right front leg to left hind leg 8,500 ohms. Connections made as above. Continuous current used. Electromotive force at the time of closing circuit, 200 volts. Time of contact through dog 2 seconds.

When the circuit closed, the St. Bernard yelped in pain. The puppy was heavier and healthier than the first dog, and received only a quarter of the voltage that the first dog had endured. The puppy continued to cry out and try to escape for several minutes, but eventually quieted down. Brown wrote that the dog was "entirely uninjured" from 200 volts of DC, but made no attempt to examine the dog closely. Having survived the experiment, the St. Bernard was subjected to another round of tests, this time with alternating current.

Third Experiment

Same dog as second experiment. Same connections. Alternating current used by introducing a circuit breaker and alternator in circuit with the dog. Electromotive force at time of closing contact through dog, 200 volts. Number of alternations, 660 per minute. Time of contact through dog, 2 seconds.

The 200-volt burst of alternating current shot into the St. Bernard and its body immediately stiffened. When the circuit was broken, the dog howled in misery and made several feeble attempts to escape. The AC jolt had clearly injured the animal, but since it had already received a shock of DC minutes before, the effects could well have been cumulative. Brown was clearly disappointed at the results, expecting that the blast of AC would have done more damage. He ordered that the puppy undergo another trial with alternating current, this time at more than three times the voltage.

Fourth Experiment

Same dog and same connections. Alternating current as in previous trial. Electromotive force at time of contact, 800 volts; number of alternations, 1,600 per minute. Length of contact, 3 seconds.

When the circuit was closed, the dog immediately became a rigid mass, looking more like a statue of a dog than a living creature. When the connection was broken, the puppy fell limp, landing hard on its side. It whimpered faintly with a single expulsion of breath and died fifteen seconds after receiving its third shock. Satisfied with the result, Brown concluded the experiments for the evening.

It's not clear whether Edison was present at Brown's first round of experiments, although the inventor eventually would view at least some of Brown's grisly work. Edison wasn't squeamish about zapping animals with electricity. Brown's dog experiments might have reminded Edison of the Rat Paralyzer he had built twenty years earlier to rid the telegraph office of rodents.

Edison also must have known that Brown's dog experiments were hardly scientific. There were no control subjects. The weight of each dog, a crucial factor in a creature's resistance to electricity, was merely estimated. The St. Bernard puppy was subjected to multiple shocks of varying voltages of both DC and AC, making it impossible to single out a fatal blow. Neither dog in the experiment was dissected, leaving Brown free to interpret the results as he wished—as proof of AC's singular deadliness.

George Westinghouse viewed Brown's sensational claims about AC with growing dismay. A letter published in the trade journal *Electrical World* signed by a Westinghouse vice president stated flatly, "The effect of the alternating current upon animal life is immensely less both as to burning and possible death than the direct current of the same volume and pressure." Westinghouse considered taking legal action against Edison, but decided that such a move would only give the anti-AC faction more publicity.

Westinghouse was appalled most of all by the growing savagery of the AC/DC battle. "The struggle for the control of the electric light and power business has never been exceeded in bitterness by any of the historical commercial controversies of a former day," Westinghouse wrote. Nonetheless, Westinghouse wasn't so high-minded that he refrained from taking a shot at the competition when he got a chance. "I have witnessed the roasting of a large piece of fresh beef by a direct continuous current of less than one hundred volts within two minutes," Westinghouse wrote in a magazine article, adding that anyone touching a live 100-volt DC wire would find it "painful beyond endurance." He argued that Brown's dog experiments didn't prove that AC was any more dangerous than DC, and noted that the voltage that entered a customer's house in the Westinghouse system was less than half the voltage of the Edison system. But Westinghouse's calmly rational arguments in support of AC's safety were drowned out by Harold Brown's carnival barker pronouncements.

Word of Brown's experiments reached a New York state commission that was interested in electricity for quite a different reason—as a means of executing criminals. In 1886, the New York State legislature had authorized a commission to investigate a more humane alternative to hanging as a method of capital punishment. Hanging had come to be seen as cruel and unusual punishment even when done properly; frequently, the hangman's ineptness produced gruesome scenes of slow strangulation and even decapitation. The commission, headed by Elbridge T. Gerry, grandson of one of the signers of the Declaration of Independence, recommended that hanging be replaced by an entirely new form of capital punishment: death by electricity (the term *electrocution* had yet to be coined). On June 4, 1888, the state legislature passed a law establishing death by electricity as the preferred method for future executions, and ordered a panel of experts to recommend how to implement the new law.

Brown's dog experiments came at just the right time for the commission. There was scant scientific information about the effects of electricity on living creatures, and no data at all on what type of current and voltage was sufficient to kill a human being. The panel's inquiries also came at a fortunate time for Brown and Edison. If the commission could be convinced that AC was so reliably deadly that it would make a splendid means of killing human beings, it would deal a devastating blow to the Westinghouse forces and the AC standard. Few families would want to welcome the executioner's current into their homes.

Two days after Brown's dog experiments, Edison invited several members of the commission to his laboratory, along with a *New York Times* reporter. The visitors were met at the Orange train station, driven to the lab, and provided with a grand tour of the Wizard's facilities. They visited Edison's phonograph room, where they were treated to a recording of a cornet solo, and then retired to one of the smaller experiment rooms. There, a Wheatstone bridge, a device used to measure electrical resistance, was waiting for them.

Edison knew that a demonstration of Brown's dog experiments would be too ghastly to show the group, especially in the presence of a newspaper reporter. Instead, he chose to stress the "scientific" aspect of the experiments. Each commission member was attached to the Wheatstone bridge and painlessly had his resistance to electricity measured. Even the *Times* reporter got in on the fun, and reported that his body had a resistance of 2,500 ohms. Every creature has a unique resistance to electricity, Edison told the group. Delivering a fatal shock of electricity was simply a matter of overcoming that resistance with a powerful enough current. Although his lab's animal experiments had just begun, Edison declared that alternating current already had shown itself to be especially deadly.

While the group was having their resistance measured, an assistant to a lawyer opposing the new execution law entered the room, accompanied by a young man whom no one recognized. After a few minutes, the stranger was unmasked as an employee of the Westinghouse Electric Light Company, a mole sent to check up on Edison's claims about the AC system.

A suitably indignant Harold Brown immediately denounced the intruder, and even the *Times* reporter thought the Westinghouse man had "overstepped the bounds of courtesy in entering the establishment of a rival when experiments might be going on." After some debate, the Westinghouse employee was permitted to stay, but the incident left Brown and Edison more convinced than ever of the identity of their true enemy. Electricity had a new set of dualities in the Industrial Age, built on the eternal ones: Positive versus negative. DC versus AC. And now, Edison versus Westinghouse.

After the commissioners left, Brown settled down to another evening of animal experiments. Edison may have suggested to Brown that his first series of tests were lacking somewhat in scientific rigor. For round two, Brown took studious measurements of each dog's weight, height, and length. He also set up a relay that illuminated eight lamps when the circuit was closed so he could better judge the length of time the current was flowing through the wire and into the dog. Dr. Frederick Peterson, a friend of Edison who used electricity in his practice to treat a variety of ailments, was on hand to dissect any dog that was killed. The second round of experiments began at 9:35 P.M., with Brown once again recording the gruesome details.

Fifth Experiment

Dog No. 3. Fox terrier bitch, young and of good vitality; weight 13½ pounds; height 13 inches; length from tip of nose to base of tail 24 inches. Resistance 6,000 ohms. Connections made as above and kept thoroughly wet. A relay was provided to close circuit upon a series of lamps as soon as circuit through dog was opened. Continuous (direct) current; Electromotive force 400 volts.

When the circuit was closed, the eight lamps flickered and the fox terrier received 400 volts of DC, half the voltage that had killed the dog of a similar size two nights before. The fox terrier let out a yelp and struggled to escape. In his notes, Brown concluded that the dog was "unhurt." Half an hour later, the dog was deemed ready for another jolt of DC, this time, at a higher voltage.

Sixth Experiment

Same dog and same connections; relay used as in previous trial. Continuous current; Electromotive force 600 volts.

The lamps flashed and the terrier cried out and struggled to break free. Once again, Brown deemed the dog to be "unhurt." This time, he waited only three minutes before subjecting the same dog to an even stronger burst of direct current.

Seventh Experiment

Same dog and same connections; relay used as in previous trial. Continuous current; Electromotive force 800 volts.

The terrier yelped as 800 volts of DC surged through its body, the same voltage that had killed the black and tan mutt two nights before. Incredibly, Brown merely noted, "Dog howled and struggled, but unhurt." Brown had no way of knowing whether the bursts of DC were causing any physiological damage, and didn't seem particularly interested in finding out. Five minutes later, Brown cranked up the voltage even higher and subjected the dog to a fourth shock.

Eighth Experiment

Same dog and same connections; relay used as in previous trial.

Continuous current; Electromotive force 1000 volts. Relay contacts apparently poor, as lamps did not start immediately, as in previous trials with same dog.

This time, the effects were apparent even to Brown. When the current hit, the dog howled and struggled violently for two minutes, and then collapsed. Believing the dog was dead, Brown ordered Dr. Peterson to perform an immediate dissection. When the doctor sliced the dog's chest open, he discovered that the animal's heart was still beating and some muscular tissue was quivering. Dr. Peterson concluded that artificial respiration could have saved the dog's life if Brown hadn't been so eager to dissect the animal. To give the botched experiment the faint whiff of science, sections of the dog's spinal cord and sciatic nerve were removed for later microscopic examination.

Brown concluded from the experiment that nothing less than 1,000 volts of DC were required to kill a 13½-pound creature. But once again, Brown's sloppy methodology didn't support his findings. The terrier had received a total of 2,800 volts over the span of about an hour; it was impossible to conclude that the final 1,000-volt shock was fatal in and of itself. By Brown's own admission, the relay contacts that attached the bare copper wire to the terrier's front and hind legs were faulty in the final experiment, enough of a variable to invalidate the entire experiment.

There was still one more dog left, and Brown decided to subject it to the "deadly" alternating current, with predictable results.

Ninth Experiment

Dog No. 4. Half-bred bulldog, strong and vigorous. Weight 40 ½ lbs.; height 21 in.; length from tip of nose to base of tail, 37 in. Connections same as above. Resistance 11,000 ohms. Alternating current; Electromotive force 800 volts; number of alternations 2,200 per minute. Time of contact through dog, 2.5 seconds.

When the circuit was closed, Brown reported that the bulldog immediately "turned to stone," reserving his most colorful description for the damnable alternating current. When the contact was broken, the dog fell limp and died within fifteen seconds. A dissection revealed that the dog's blood "had a peculiar dark, thin fluid appearance," a curious observation Brown made no attempt to explain further.

Brown was pleased that 800 volts of AC had dispatched the bulldog so quickly, although the same voltage of DC had killed the black and tan mutt in the first experiment. Two nights later, Brown conducted a third round of experiments, hoping to show DC in a more favorable light.

Tenth Experiment

Half-bred shepherd dog; strong and in good condition; weight 50 lbs.; height 23½ in.; length from tip of the nose to base of the tail, 39 in. Connections same as above. Resistance 6,000 ohms; continuous current, Electromotive force 1,000 volts.

When the circuit was closed, the shepherd yelped once, but according to Brown was "entirely unhurt." Satisfied that the dog had survived a 1,000-volt burst of DC when 800 volts of DC had killed the first dog, Brown relentlessly increased the voltage for five consecutive trials on the same dog, each just minutes apart.

Eleventh Experiment

Same dog and same connections. Continuous current; Electromotive force 1,100 volts.

Current closed at 9:44 p.m. Respiration fell to 72 and dog unhurt. Dog yelped when circuit was closed, but wagged his tail as Dr. Peterson counted respiration.

Twelfth Experiment

Same dog and same connections. Continuous current; electromotive force 1,200 volts.

Circuit closed at 9:46 p.m. Dog yelped as circuit was closed, but still unhurt. Respiration, 72.

Thirteenth Experiment

Same dog and same connections. Continuous current; Electromotive force 1,300 volts.

Circuit closed at 9:51 p.m. Dog yelped as circuit was closed, but still unhurt. Respiration, 60.

Fourteenth Experiment

Same dog and same connections. Electromotive force 1,400 volts.

Circuit closed at 9:53 p.m. Dog yelped slightly as the circuit was closed, but still unhurt. Respiration 72 (irregular).

Fifteenth Experiment

Same dog and same connections. Continuous current; Electromotive force 1,420 volts, the utmost capacity of dynamo at present speed; all resistance removed from field circuit. Circuit closed at 9:58 p.m. Dog yelped, but unhurt. Respiration, 72. Dog removed from box and found to be entirely uninjured. No signs of paralysis in either sensory or motor nerves.

Even some of Edison's hard-bitten lab assistants were unnerved by the sight of the dog being subjected to six successive shocks of DC. Brown, however, wasn't through yet. All of the experiments on the shepherd were with a single, instantaneous closing and opening of a circuit. Now Brown wanted to see if holding the relay armature for two seconds after the circuit was closed would make any difference. The workers reluctantly returned the dog to his box.

Sixteenth Experiment

Same dog and same connections. Armature held for 2½ seconds after circuit through dog was closed. Continuous current; electromotive force 1,200 volts. Circuit closed at 10:30 p.m.

The shepherd cried out and struggled to escape as the current was applied and the armature was held for two and a half seconds. When the circuit was broken, there was silence and everyone peered expectantly in the box. Amazingly, the dog was still alive, though badly shaken by the ordeal. One of Edison's engineers scooped the dog out of the box, lifted him into his arms and declared that he was adopting him on the spot. The engineer named the dog Ajax, after the Greek hero who had defied lightning. Perhaps the engineer was unaware that Ajax was eventually struck dead by lightning for his insolence.

There was no reason for further experiments, if there had been reason for any of them. Brown had "proven" that AC was more deadly than DC simply by making the facts fit his original premise. Still, Brown pressed on. Over the next two weeks, he conducted eleven more dog experiments using alternating current, all ultimately resulting in the dog's death. The later experiments weren't conducted any more carefully. One dog, alarmed at being confined in a small box, emptied his bladder shortly before the circuit was closed. When the current surged through the dog, the puddle of urine acted as a conductor. Not surprisingly, the dog "screamed and howled loudly during the time of closing and made violent effort to break loose for about one minute after circuit was opened." Brown's solution was simply to place the dog in a dry box and resume the experiment.

Before he was finished, Brown experimented on forty-four dogs at the Edison lab, torturing them all and killing all but a handful. Brown showed no remorse over the suffering he inflicted, nor any scruples about interpreting the results.

"I determined, to the satisfaction of Mr. Edison and other prominent scientists, the exact pressure required to produce death with the continuous and with the alternating current," Brown declared. "The result proved that the alternating current at 160 volts, or less than one-sixth the pressure used for electric lighting by the Westinghouse and other alternating current companies, was instantly fatal, while with the continuous current, no injury whatsoever resulted from a pressure of 1,420 volts."

Of course, not even Brown's self-serving experimental notes supported this conclusion. In Brown's first experiment, he had killed a dog with 800 volts of DC; to say that no injury whatsoever resulted from a pressure of 1,420 volts of DC was simply untrue. Edison must have known that Brown's experiments were pseudoscience, but they provided just the sort of conclusions he had been looking for in his increasingly bitter fight against AC.

Brown's findings were met with hoots of derision from the Westinghouse forces. How could Brown expect anyone to believe results drawn from experiments that were closed to the public and performed by an electrician with a long-standing interest in promoting DC over AC? Once again, the prickly Brown took the attacks as an assault on his reputation, and announced plans for a public demonstration to prove to the world that his conclusions about the deadliness of AC were scientifically sound.

On the afternoon of July 30, close to eight hundred curious onlookers packed into a lecture room at the Columbia University School of Mines in New York City to view Brown's latest demonstration. Brown had sent invitations to representatives of all the electric light companies, and the Westinghouse group had shown up in force. The New York City Board of Electrical Control was also on hand, along with journalists from the leading New York newspapers. Edison did not attend, but loaned the services of his chief electrician, along with electrical equipment from his lab. Edison was willing to lend his support to Brown, but not his name.

Just two months before, Harold Brown had been an anonymous light salesman scraping out a living as a self-proclaimed electrical expert. Now he was a prominent figure in one of the great controversies of his day, a force to be reckoned with. Brown strode to the front of the lecture hall and began to speak in magisterial tones befitting a true Man of Science.

"Gentlemen, it is only by my sense of right that I have been drawn into this controversy," Brown began. "I represent no company and have no financial or commercial interest."

A wave of derisive laughter rippled from the AC backers. Brown pressed on.

"I do not propose to present a scientific paper to you this afternoon, but will simply give you a few samples of the experiments I have been engaged in for a considerable time," Brown continued. "I have proved by repeated experiments that a living creature could stand shock from a continuous current much better from an alternating current. I have applied a current of 1,410 volts to a dog without fatal result, and I have repeatedly killed dogs with from 500 to 800 volts of alternating current. Those advocates of the alternating system who claimed to have withstood a shock of 1,000 volts of alternating current without injury must have worn lightning rods."

Again, a jeer wafted up from the AC crowd. The packed room was growing hotter and Brown quickly wrapped up his comments and moved on to the demonstration. Brown stepped offstage and returned with a large seventy-six-pound Newfoundland dog in tow. With the help of several assistants, Brown attached a muzzle to the dog, placed it in a wire cage and tied it down with a rope. He carefully measured the dog's resistance, and announced it to the crowd as 10,300 ohms. This was more than four times the resistance of a reporter from the *New York Times*, according to Brown's measurements just weeks before. Still, the show went on; wires leading from a generator were attached to the dog's front and hind legs.

"My first experiment will be with the continuous current," Brown intoned. "I shall apply a continuous current with an electromotive force of 300 volts." The restive crowd was silent now. Brown closed the circuit and the current surged into the dog, which let out a frightened yelp. Some in the crowd flinched.

"As you can see, the dog is unhurt," Brown announced. "I shall now increase the force to 400 volts."

This time, the dog's cry was more piercing as it made a desperate effort to escape. Audience members shifted uncomfortably in their seats.

"And now 700 volts."

The dog screamed in pain and thrashed violently in its cage. Its movements became so frantic that the dog broke free of its muzzle and snapped the rope that was holding it down. A murmur swept through the crowd. Brown had the dog tied down again for yet another test.

"One thousand volts."

The circuit closed and the Newfoundland cried out, its entire body contorting with pain. Some in the audience turned their heads. Brown moved in quickly to deliver the crowning blow.

"This dog will have less trouble when we try the alternating current," Brown announced. And then with a snigger he added, "As these gentlemen say, we shall make him feel better."

The circuit was closed and a 330-volt burst of alternating current surged into the hapless Newfoundland. This time there were no cries. The dog collapsed like a rag doll.

Immediately, there were indignant cries from the AC backers. The Newfoundland had clearly been close to death from successive jolts of DC; the final burst of AC merely finished the job, they argued. Brown stepped off the platform and returned with another dog, which he said would be subjected solely to alternating current.

As Brown prepared to place the new dog in the cage, a stranger suddenly stepped onto the stage. He flashed a badge, identifying himself as Agent Haukinson of the SPCA, and ordered that the experiment be halted immediately. A crestfallen Brown reluctantly led the dog offstage, but the crowd was now in full fury. One alternating current supporter stood up and shouted that if Brown truly believed DC was harmless, he'd have no problem putting his convictions to the test. The man proposed to subject a member of his company to 1,000 volts of AC if Brown would agree to take the same voltage of DC.

This suggestion brought a roar of approval from the crowd, delighted at the prospect of having Brown get a taste of his own medicine. Brown, however, declined to participate in the proposed electric duel, to the manifest disappointment of many in the audience.

"I wish this experiment had not been interrupted," Brown declared as the meeting broke up in chaos. "I have enough dogs to satisfy the most skeptical. The only places where alternating current should be used are the dog pound, the slaughterhouse, and the state prison!"

The demonstration had been an utter debacle. It proved nothing besides Brown's cruelty and cowardice. Even the vainglorious Brown sensed that his exhibition had been a failure, and he quickly convened a second round of experiments at Columbia four days later. This time the demonstration was closed to the public, and the AC forces flatly refused to attend anyway. Several physicians advising the New York legislature about the new execution law were on hand to watch Brown send three more dogs to their death with bursts of alternating current.

A report signed by Brown and the attending physicians came to a rather dubious conclusion considering what Brown had seen from his own experiments: "All of the physicians present expressed the opinion that a dog had a higher vitality than a man and that therefore a current which killed a dog would be fatal to a man under the same conditions. It was their opinion that all of these deaths were painless, as the nerves were probably destroyed in less time than that required to transmit the impression to the brain of the subject."

One of the signing physicians was Dr. Frederick Peterson, who had assisted Brown in his second round of dog experiments at the Edison lab. Soon after the Columbia experiments, Dr. Peterson was appointed chairman of the Medico-Legal Society committee that would make detailed recommendations about how best to implement New York's new execution law.

Brown could scarcely believe his good fortune. He had launched his anti-AC campaign with modest goals, hoping to sway public opinion in the debate over electrical safety while drumming up some new business for himself. Now, Brown had a powerful ally on the committee that would decide what kind of current—AC or DC—would be used to kill human beings. The world was starting to view Harold Brown the way he had always seen himself—as an expert.

8

OLD SPARKY

Now it was total war. What began as an ordinary skirmish between competing technical standards had deteriorated into a grotesque campaign of lies and fear mongering. The high stakes had brought out the worst in practically everyone. The winners in the AC/DC battle stood to control the electricity market for decades to come; the losers would be forced to retool their entire operation at great expense or risk going out of business. The mysterious nature of electricity only made it easier to make sensational public claims. Reasoned arguments in favor of direct or alternating current were no match for appeals to fear, the deeply rooted dread of lightning that humans had carried around for tens of thousands of years.

Harold Brown instinctively played to these subterranean fears, using the tools of modern science to dredge up terrors from the Dark Ages. In the fall of 1888, Brown began to compile an inventory of people supposedly killed or crippled by electricity, a list that, not surprisingly, left no doubt as to which current posed the greater threat. In a letter published in the journal *Electrical World*, Brown wrote that his research showed "that the number of victims of the alternating current in this country is already far in excess of its proper proportion. . . . Each additional homicide caused by poor insulation, inadequate testing for grounds or too high pressure, will be an argument in favor of legislative prohibition instead of wise legislative regulation, which not even an alternating current trust of twenty millions can overcome."

Only months before, Brown had been saying that the AC forces were backing an unsafe standard largely out of ignorance. Now he charged that the public was being imperiled by a $20 million "trust" of callous businessmen that valued profit over human life. To counter Brown's increasingly shrill claims, AC backers called on Dr. Peter H. Van der Weyde, a well-regarded electrical expert, to write a paper attacking Brown's animal experiments as pseudo-science. Members of the National Electric Light Association, an industry group largely controlled by alternating current companies, greeted the paper enthusiastically. The association unanimously adopted a resolution stating, "that it is our conviction that there is no difference in the danger attending the use of continuous or alternating currents and that both may be so transformed before being used as to render them perfectly harmless and tractable means of distributing electric power for our cities."

AC was making rapid gains in the marketplace, but its backers were at a distinct disadvantage when it came to competing in the court of public opinion. The alternating current companies were, for the most part, content to refute Brown's outlandish claims with technical papers that argued that neither AC nor DC was inherently more dangerous. Harold Brown had no such scruples. Science was merely a club one used to beat an opponent.

Brown set his sights on convincing New York's Medico-Legal Society to recommend AC as the most effective means of executing criminals. One of the criticisms of Brown's dog experiments was that the animals weighed only 20 or 30 pounds and that the effect of electricity on, say, a 180-pound adult human being might be very different. To counter the weight argument, Brown reached for his favorite club.

On December 5, Brown held another round of animal experiments at the Edison lab, inviting the Medico-Legal Society, Elbridge Gerry, and members of the press. This time, Edison was on hand to view the experiments, a measure of how important Brown's public demonstrations had become to the inventor and to the survival of Edison's DC empire.

Brown brought along two 145-pound calves and a "strong and vigorous" horse weighing 1,230 pounds. The calves were the first subjects. Brown connected two electrodes to the first calf, placing one on the spine between the shoulders and the other on the forehead, directly between the eyes. The electrodes were wrapped in sponges that had been soaked in a salt solution, and the wiring of the circuit was more complex than in previous demonstrations. Brown fashioned a relay that stopped the flow of current whenever a ground connection was made, so that the electrodes could be removed from the animal without having to worry whether the current was still flowing. Brown couldn't resist mentioning that the same safety apparatus could easily be installed on arc lights that would make it almost impossible for accidents to occur, and that he just happened to make these devices.

Brown also decided to close the circuit in a more dramatic fashion than in previous experiments. Wires from the two electrodes led to a metal plate that rested on the floor; the circuit would be completed by banging the plate sharply with a hammer. Brown did the honors himself, bringing the hammer down on the first calf with a 770-volt blast of alternating current. The second calf was dispatched in short order with 750 volts of AC. The deaths were much tidier this time; each animal keeled over after about ten seconds. The salt-soaked sponges on the electrodes were clearly an advance; the saline was an excellent electrical conductor. When Brown removed the electrode from the calf's forehead, there was a scorch mark the size of a silver dollar between its vacant eyes.

The horse was next. On a suggestion from Edison, the electrodes were connected to the horse's forelegs. Brown's hammer came down on the metal plate. The horse stared blankly back at the group, completely unaffected. A flustered Brown quickly had the wiring inspected; a converter was deemed defective and replaced. Brown again picked up the hammer and brought it down on the metal plate. Clang! Nothing—the horse stood motionless. Brown was at a loss, and began to strike the metal plate repeatedly. Steam rose from the sponge-covered electrodes on the horse's legs, but the horse remained uninjured.

Brown sheepishly halted the experiment and the electrodes were removed and reattached to the horse's forelegs. This time, Brown left nothing to chance. Clang! The hammer came down on the metal plate and 700 volts of alternating current surged through the horse for twenty-five seconds, the longest sustained burst of electricity in any of Brown's experiments. When the circuit finally was broken, the horse fell over on its side and died immediately. Brown had the horse photographed before and after the current was applied to prevent critics from claiming that the animal was nearly dead before the fatal blow.

Despite the troubles with the horse, the Medico-Legal Society committee was impressed with Brown's demonstration of the killing power of electricity. The next day's *New York Times* account of the experiment concluded by observing "alternating current will undoubtedly drive the hangmen out of business in this state." Once again, Brown had stacked the deck by subjecting the animals only to alternating current; a similar blast of direct current might well have had the same results. And the presence of Edison at the proceedings lent the experiment an air of legitimacy.

A week later, the Medico-Legal Society held its annual meeting at New York's Fifth Avenue Hotel, with Harold Brown in attendance. The main business was to consider the committee's report about how best to use electricity to execute criminals. Drawing its information from Brown's lab notes, the committee reported that experiments on twenty-four dogs, two calves, and a horse had shown that an alternating current as low as 160 volts could kill a living creature, while a much higher voltage of direct current was necessary to produce a fatal effect.

Brown's experiments, however, were far too crude to have supplied meaningful data about the lethality of electricity. As it turns out, voltage isn't the only factor in a current's deadliness; the current's frequency, duration, and rate of flow also play key roles. A high voltage paired with a high current volume is usually lethal, but the same high voltage and a low current flow might not be deadly. The rate of flow of an electric charge was often what killed people—

rather than the voltage, or the pressure under which the electricity flows. Modern-day cardiac defibrillators, for instance, deliver a high-voltage shock to the heart, as much as 1,800 volts, but the shock is not deadly because the volume delivered is extremely low.

The committee was ignorant of the finer points of electricity, and deferred to Brown's supposed expertise. Death by AC "was without a struggle," the committee reported, while killing with direct current was accompanied by "howling and struggling." The committee therefore recommended that AC be adopted as the executioner's current.

The committee made a series of detailed recommendations about how the deadly current should be administered. An early idea that the prisoner be immersed in water to act as a conductor was rejected; so was a scheme to place large metal plates upon the condemned man's body. "It is well known that if metal be directly in contact with the skin during the passage of an electric current, burns and lacerations are apt to be produced," the committee reported. There had been talk of having the prisoner be in a standing position when the current was applied, but the committee rejected that proposal as well. "There are so many histories of unseemly struggles and contortions on the part of criminals executed by the old methods, that the necessity of some bodily restraint is evident," the committee reported. "In our opinion the recumbent or sitting position is best adapted to our purposes."

The panel recommended that the condemned man be placed "in a chair especially constructed for the purpose," the origin of what would become the electric chair. The prisoner would be strapped securely to the chair with two leather buckles, one wrapped around his midsection and the other around his forehead. One electrode would be placed on the prisoner's spine between the shoulder blades; the other would be attached to a helmet, which was fastened to the back of the chair. The electrodes would be made of metal, four inches in diameter and covered with thick layers of sponge. The sponges, as well as the skin and hair at the points of contact, would be thoroughly wet with a solution of zinc sulfate. A dynamo capable of producing at least 3,000 volts of alternating current would supply the power, with the current allowed to flow into the prisoner from fifteen to thirty seconds "to ensure death." The committee's recommendations were adopted unanimously, and the members repaired to the Palette Club on 24th Street where they held an elaborate banquet.

Harold Brown tucked into his meal a satisfied man, having gotten everything he could have hoped for. A seemingly objective group of medical professionals had ruled that alternating current was far more dangerous than direct current, and recommended that it be used to execute criminals. The full committee, however, had only witnessed Brown's final experiment, which used AC to kill subjects. No one had viewed Brown's earlier tests at the Edison lab that showed DC to be just as deadly as AC.

George Westinghouse was horrified when he heard the news. Harold Brown's scurrilous charges now had official sanction; the State of New York had declared that alternating current was singularly effective as a means of killing human beings. Westinghouse wrote a hasty public letter, which was printed in several New York newspapers on December 13. In the letter, Westinghouse attacked Brown as being in the pay of Edison interests and charged that his experiments were scientifically invalid. "The method of applying the current used in these experiments was carefully selected for the purpose of producing the most startling effects with the smallest expenditure of current," Westinghouse fumed. Westinghouse claimed that Brown's experiments were a desperate attempt by the Edison forces to defame a technology that was defeating them in the marketplace.

The prickly Brown rose to the challenge, firing off a letter of his own, which ran in the *New York Times* the following week. Brown denied "that I am now or have ever been in the employ of Mr. Edison or any of the Edison companies," although he declined to mention that his main source of income was in selling and servicing DC systems. Brown maintained that his experiments had proven beyond doubt the danger of the "death-dealing alternating current" and charged that Westinghouse continued to defend AC purely for business reasons. Brown concluded his letter with an astonishing dare:

"I therefore challenge Mr. Westinghouse to meet me in the presence of competent electrical experts and take through the body the alternating current while I take through mine a continuous current. The alternating current must have not less than 300 alternations per second. We will commence with 100 volts and will gradually increase the pressure 50 volts at a time, I leading with each increase, until either one or the other has cried enough and publicly admits his error."

It had come to this: AC and DC squaring off at high noon. It was only eight years after the famous gunfight at the OK Corral in Arizona. Brown was proposing a similar public duel, waged not with pistols but with electricity. Brown had gotten the idea for the electric duel from his public demonstration at Columbia University, when the angry AC backer challenged Brown to take 1,000 volts of DC into his body. At the time, Brown had refused the dare; now he was willing to challenge Westinghouse to a lower-voltage duel, suggesting he had come up with a way to rig the test to his satisfaction. Harold Brown liked experiments with reliable results.

George Westinghouse could only shake his head in disgust. Appearing on the same stage as Brown would only legitimize his claims and call attention to the connection between AC and execution. Westinghouse met Brown's latest challenge with silence; there was nothing to be gained by even offering a reply. Tesla, for his part, stayed out of the ugly battle altogether; he was preoccupied with other matters. After his year in Pittsburgh, Tesla had moved back to New York and opened a small office on Grand Street. He was also applying to become a U.S. citizen.

Brown continued to bait Westinghouse. "[Westinghouse] is willing to risk the lives of the public by stringing his death-carrying wires recklessly through our streets, but he knows too much to place his own life at the mercy of the deadly alternating current," Brown told reporters. "I am told, although living within reach of an alternating current station, he has preferred to use the continuous current in his own house."

Brown's increasingly desperate pronouncements were, in a way, evidence that DC was losing the battle. In October 1888 alone, Westinghouse received orders to power 45,000 lights on the AC system, about what the Edison companies had sold for the entire year on the DC system. The order included a 25,000-lamp consignment for London, which had much stricter electrical safety laws than New York. By 1890, Westinghouse Electric's revenues soared to $4 million. Harold Brown could kill all the animals he wanted; the market was voting for AC.

No one knew this better than Edison, who always kept a close watch on sales figures and market share of his inventions. Edison wasn't in the habit of losing, and the idea of defeat in such a large enterprise as electricity only stoked his competitive nature. He focused his anger on the man who stood to gain most from his defeat, George Westinghouse.

In early 1889, E.D. Adams, a partner in an investment bank and a close friend of Westinghouse, approached Edison with a peace offer. Adams was traveling to Pittsburgh to meet Westinghouse and suggested that Edison come along and make amends. After all, the two men were both inventors and electrical pioneers in their own right. They had brought light and power to thousands of people and given birth to an entirely new industry. Beneath the surface rivalry, the two men had much in common, inventors who thought big. Instead of fighting each other, why not agree to a truce, or perhaps even join forces?

Edison answered with a telegram dripping with scorn: "Am very well aware of his resources and plant and his methods of doing business lately are such that the man has gone crazy over sudden accession of wealth or something unknown to me and is flying a kite that will land him in the mud sooner or later."

The war of the currents raged on.

On January 1, 1889, New York's execution law had gone into effect, the world's first statute to specify that electricity be used for capital punishment. The law was widely hailed as an enlightened step forward, a civilized solution to the long-standing problem of how to put a criminal to death humanely. No longer would condemned men suffer needlessly at the hands of inept hangmen; the proposed "death chair" would quickly and painlessly send murderers to their Maker. The *New York World* hailed the chair as "a highly scientific device for electrical executions," and news articles made the apparatus sound more like a carefully constructed medical device than a jury-rigged killing machine.

To Edison and Brown, the new execution law offered the best means of dealing AC a decisive blow. Edison was philosophically opposed to the death penalty, but he didn't let scruples get in the way of business. In March 1889, Brown made a final series of animal tests, this time dispatching four dogs, four calves, and a horse with 800–1,000 volts of alternating current. Edison once again lent use of his laboratory to conduct the experiments, which were viewed by members of the Medico-Legal Society as well as the medical superintendent of Auburn state prison.

Brown's last round of experiments proved, if nothing else, that he was getting better at killing animals with electricity. The electrodes were even more carefully designed this time, constructed of a copper wire coil wrapped in layers of cotton that had been soaked in zinc sulfate. High voltages were used and the current was applied for longer periods of time, as much as eighteen seconds. The results were predictable, the way Harold Brown liked them. All nine animals died with little struggle.

Already, the committee was convinced that alternating current would be an effective and humane means of capital punishment. Brown further persuaded the committee that constructing a special apparatus to power a death chair would be too costly. An off-the-shelf commercial AC dynamo would be cheaper and more reliable. In fact, Harold Brown knew just the model.

Brown met with Austin Lathrop, the superintendent of New York prisons, who placed Brown in charge of purchasing the dynamos that would power the new electric chair. The Westinghouse companies flatly refused to do business with Brown or the prisons, but Brown was not so easily deflected. Through an intermediary, Brown purchased three Westinghouse AC generators that were shipped to New York. Brown made sure that the generators produced the same voltage as the standard Westinghouse 650-light system already in use. That way, the AC sent to the death chair would be identical to the current that flowed into thousands of homes and offices. The identification of the Westinghouse system with death would be complete.

It would be more than a year before the death chair was called upon to claim its first victim. Edison and Brown used the time to press their momentary advantage and stir up more bad publicity for alternating current. A *New York World* reporter asked Edison, "What about the rumor that some of your batteries were sold to the State of New York to use in the execution of criminals?" Edison smiled and replied, "Oh, that was the Westinghouse engines, not mine."

Brown peppered the New York newspapers with accounts of the latest horrors caused by the deadly alternating current. In July 1889, Brown wrote in a published letter that ten people had recently been killed by AC, warning that the list was growing daily. According to Brown's figures, deaths from alternating current had jumped from just three in 1887 to twenty-four in 1888–89.

George Westinghouse dispatched several men to see if Brown's figures bore any relation to the truth. The Westinghouse investigators found that of the nearly thirty deaths Brown attributed to alternating current, only one could be confirmed as caused by AC. In twelve of the supposed AC deaths, there were no Westinghouse plants in the city at the time of the accident. In sixteen cases, arc lighting—which ran on the DC system—was the culprit. Furthermore, the overwhelming majority of those killed by electricity were electrical linemen installing or servicing power lines. The deaths Brown cited were more an argument for safer working conditions in the electrical industry than for limiting the spread of AC. The information gathered by Westinghouse's investigators was sent to all the company's sales agents to reassure customers.

Brown's spectacular claims, however, made good copy. He was now a fixture in the New York papers, variously described as "a prominent electrical engineer" or "New York State expert on electrical execution." It was heady stuff, and Brown took full advantage of his growing reputation. He began to present "serious" papers to medical and legal groups, filled not only with dire warnings about AC but also with visions of a grand future powered by safe, reliable direct current. In a speech given to the International Medical Jurisprudence Congress in New York, Brown described the coming DC-powered utopia: "The air will no longer be polluted with smoke, for one immense station provided with triple or quadruple expansion engines and furnaces in which combustion is complete will supply heat, light, power, and motion. The consequent addition to human health, comfort, and length of life by the banishment of dirt and noise will be enormous. Electrical disinfection and sewage purification are already in use and since we can command immense volumes of electricity, it is not improbable that a better understanding of the laws of meteorology will enable us at least partially to control the weather, and thus avoid the evil effects of severe changes and extreme temperatures."

Naturally, such a future would be impossible unless alternating current was regulated out of business. "Earth and air are filled with wires, many of which may be charged with swift and invisible death," Brown declared. "It is clearly the physician's duty to point out the dangerous currents and it remains for the lawyer to secure wise legislative action preventing the adoption of systems or apparatuses which needlessly jeopardize human life or health."

According to Brown, alternating current companies were being allowed "to enmesh our cities with wires carrying death-dealing currents—currents which can escape and produce death through any known insulation." Special legislation limiting AC voltages was the only answer. Without such safeguards, electricity would never achieve its fantastic potential.

Brown brought his arguments to an even wider audience when he and Edison wrote companion articles in the November 1889 issue of the *North American Review*, an influential magazine of the day. Brown's article, "The New Instrument of Execution," recounted his work on the death chair and the special killing power of AC;

Edison's piece, "The Dangers of Electric Lighting," was a plainspoken denunciation of alternating current. (Other articles in the same issue included, "An English View of the Civil War," "The Hopes of the Democratic Party," "The Future of Fiction," and "Are Telegraph Rates too High?") Edison had a long history with the *North American Review.* As a young telegrapher in Louisville, he paid $2 for a set of twenty volumes of the publication, which frequently featured scientific articles. Later, he wrote articles for the magazine describing his latest inventions.

Edison had called in some favors to get the *North American Review* to publish two long articles attacking alternating current in the same issue. Brown had never published a magazine article, but his association with Edison was credential enough. Brown's article set forth the conclusions he drew from his animal tests, although tellingly, he omitted any description of the experiments themselves. "My experiments showed that the greater the number of alternations per second, or the longer time of contact with the subject, the less was the pressure required to destroy life. . . . The main effect appears in violent vibrations of fluids and tissues, delivering tremendous blows within the vital organs. This is undoubtedly the secret of the life-destroying power possessed by alternating current."

According to Brown, the only appropriate use of alternating current was to kill. Even though an AC-powered execution chair had yet to be constructed, Brown was supremely confident about its performance. Describing an execution in the not-too-distant future, Brown wrote, "The deputy-sheriff closes the switch. Respiration and heart-action (of the prisoner) instantly cease, and electricity, with a velocity equaling that of light, destroys life before nerve-sensation, at a speed of only one hundred and eighty feet per second, can reach the brain. There is a stiffening of the muscles, which gradually relax after five seconds have passed; but there is no struggle and no sound. The majesty of the law has been vindicated, but no physical pain has been caused."

Edison's companion article, "The Dangers of Electric Lighting," picked up where Brown's left off. Nothing Edison wrote on the subject of AC and DC would be more impassioned, nor more distorted by his own self-interest. Edison's article began in a revealingly defensive tone. He was clearly embarrassed by Brown's crude experiments, but was unwilling to reject the useful conclusions to be drawn from them. "The public would scarcely be interested in the details leading up to the position taken by myself and the conclusions to which I have come," Edison wrote of Brown's experiments. "But I may say that I have not failed to seek practical demonstration in support of such facts as have been developed, and I have taken life—not human life—in the belief and full consciousness that the end justified the means."

This would stand as Edison's only public statement about Brown's experiments, a half-hearted apology that only hinted at the savage excesses performed in the name of science. Such means were justified by AC's unique danger, Edison argued; the AC death count was now put at one hundred victims. Putting the AC wires underground, as some were suggesting, wouldn't do any good, Edison said. The inventor told a harrowing story in which the underground conductors of an Edison power line under Wall Street accidentally became crossed. Even though the DC line ran at the "safe" power of 110 volts, it "melted not only the wires, but several feet of iron tubing in which they were encased, and reduced the paving-stones within a radius of three or four feet to a molten mass." What would the effect have been, Edison asked, if the pressure were 2,000 volts of AC?

"There is no plea which will justify the use of high-tension and alternating currents, either in a scientific or commercial sense," Edison continued. "They are employed solely to reduce investment in copper wire and real estate. . . . When an alternating current of fifteen volts is applied to a human being in the most effective manner, the effect upon the nerve system is so violent and the pain produced so great that it is absolutely impossible for any one to stand it."

The only answer, declared Edison, was to pass a law restricting electrical pressures to 600–700 volts. (Edison had raised his earlier proposed limit of 300 volts because he wanted to leave open the option of increasing the voltage of his DC system to offset rising copper wire prices.) A 600-volt limit would still effectively legislate the AC companies out of business. Edison rejected alternating current systems not only on account of their danger, "but because of their general unreliability and unsuitability for any general system of distribution."

Edison revealed that his own company, over his vigorous protests, had purchased the patents for a complete AC system. "Up to the present time I have succeeded in inducing them not to offer this system to the public, nor will they ever do so with my consent. My personal desire would be to prohibit entirely the use of alternating currents. They are as unnecessary as they are dangerous."

In fact, Edison's company had purchased an AC system based on a Hungarian design that was being operated successfully in several cities in Europe. The company purchased the AC patents in 1886, and a report by one of Edison's top electricians strongly urged him to adopt the AC standard because of its economy in long-distance transmission.

Even at this late date, Edison could have shifted some of his company's resources to the AC standard and quickly made up lost ground on Westinghouse. Edison had built up a manufacturing and marketing organization second to none in the electric industry; a strong move into AC would have made the Edison companies hard to beat. Furthermore, Edison's investment in DC didn't have to go entirely to waste. Edison could have adopted a hybrid system that would transmit power over long distances by alternating current and then convert the power to direct current for use in homes and offices.

But Edison stubbornly refused to budge; the AC patents his company had purchased were allowed to lapse. Edison had been handed his best chance of defeating Westinghouse and petulantly threw it away; he had become a defender of the old order rather than someone who challenged it. Edison had sunk too much money and—always more important for him—invested too much of his reputation on the direct current system. The louder the clamor for AC, the more Edison turned his famously deaf ear to the din. *What I have needed to hear, I have heard.*

It was a rare failure of imagination on Edison's part. Edison's direct current distribution system was the sort of plan that came naturally to someone who grew up in a small town. Under the Edison system, every hamlet in the country would have its own self-contained DC power station, serving local needs, like the village blacksmith or butcher. The Westinghouse AC system, by contrast, was conceived on a national scale, more like the railroads with which George Westinghouse was so familiar—a large network of long-distance routes.

By now, Edison's only hope of defeating AC was to make people afraid of it. Brown's experiments had seen to it that AC was chosen as the executioner's current. Now, all that was left was to select the death chair's first victim.

It came in the person of William Kemmler, an illiterate, alcoholic vegetable peddler from Buffalo. On the morning of March 29, 1889, Kemmler drunkenly accused his common-law wife, Tillie Ziegler, of planning to leave him. A bitter argument ensued, and Kemmler picked up a hatchet and struck Ziegler until there was no more arguing. Kemmler immediately walked to his neighbor's house and confessed.

"I killed her," Kemmler said. "I had to do it. I meant to. I killed her and I'll take the rope for it."

But the rope was soon to be as dead as Tillie Ziegler. Six weeks after the killing, Kemmler was convicted of first-degree murder and was sentenced to die in Auburn prison. As the first criminal sentenced to death in New York State in 1889, Kemmler would be the first to be killed by electricity.

W. Bourke Cockran, a prominent and high-priced lawyer of the day, took on Kemmler's case. Cockran assured reporters that he had taken the case in the interests of humanity, failing to mention that his own interests were being taken care of by George Westinghouse. Fearing that the Kemmler execution could hurt his company's standing, Westinghouse quietly handled Cockran's fee, estimated to be as much as $100,000.

Cockran managed to delay Kemmler's execution for more than a year by arguing that death by electricity would violate the Eighth Amendment's prohibition against cruel and unusual punishment. In

July 1889, Cockran initiated proceedings against Charles Durston, the warden of the state prison in Auburn. A state judge conducted hearings to examine Cockran's claims; Thomas Edison and Harold Brown were among those who were called to testify.

Cockran's main argument in Kemmler's behalf was that electricity was far too unpredictable to be a reliable or humane means of execution. The deadly effects of electricity were little understood, and there was great variation in how much voltage could be safely taken into the body. During the hearing, Cockran called several witnesses who testified to having received massive bursts of electricity, yet had walked away unharmed. Dr. Landon Carter Gray, a New York physician and medical expert, testified that the effect of electricity on the human body was far too unpredictable for the death chair to be a reliable means of capital punishment.

"Men have been killed by electricity, it is true, both in the artificial form and by lightning, but other men have been struck by thunderbolts or come in contact with large artificial currents without injury," Gray testified. "To attempt to put a person to death with our present knowledge of the fatal effects of electricity might lead to horrible scenes and even great fraud. . . . If the current were not powerful enough or if the resistance of the criminal was very great, he might merely be tortured and racked and suffer the agonies of death without its relief."

Harold Brown took the stand on July 8, recounting the results of his animal experiments, which he said proved that AC could kill a human being quickly and painlessly. Kemmler's lawyers hammered away at Brown's credibility, noting his lack of formal training in electricity or medicine. On the witness stand, Brown held firm to his expert status:

Q: And you have got no medical knowledge?
Brown: Except in an electro-medical way.
Q: Describe what you mean by an electro-medical manner.
Brown: Except in a general way, except as to the application of electricity to the human body.

Q: That is to say, you have seen experiments of the application of electricity to the human body?
Brown: Yes sir, and I have taken part in them.

In describing his animal experiments, Brown did his best to put a scientific gloss on his tests. But lawyers managed to get Brown to admit that some dogs were subjected to multiple electric shocks "if our supply of dogs was limited, and if the gentlemen present had come several miles to attend the experiments."

Kemmler's lawyers noted Brown's strong ties to Edison and charged that his experiments were meant to serve the inventor's commercial interests. Brown replied that Edison was merely "a personal acquaintance," and haughtily denied his experiments were motivated by anything other than public safety. Astonishingly, Brown testified that he was only vaguely aware of a conflict between Edison and Westinghouse:

Q: There is a contest between the Westinghouse Electric Light Company and the Edison Electric Light Company as to the use of these incandescent burners?
Brown: I understand so.
Q: And there is considerable feeling between the two corporations?
Brown: Of that I cannot say.
Q: Don't you know anything about it at all?
Brown: Not from actual knowledge.

To refresh Brown's memory, Kemmler's lawyers produced one of Brown's own pamphlets, "The Comparative Danger to Life of the Alternating and Continuous Electrical Currents." The back of the pamphlet featured Brown's challenge to George Westinghouse to fight a public duel with electricity.

Thomas Edison took the stand on July 23 and quickly dismissed the defense's arguments about electricity as "nonsense." As long as sufficient voltages were used, Edison said, the electrified chair would do its work quickly and painlessly. Place the criminal's hands in jars filled with a solution of potash and water, Edison suggested, and deliver a 1,000-volt burst of alternating current to the man's head and spine. Edison said he was certain of the results; he had seen experiments on animals for himself that proved the deadliness of the alternating current. When asked to describe the experiments, Edison cagily replied that he'd rather have his chief electrician testify on that point.

On October 9, the court denied Kemmler's appeal, clearing the way for the murderer to be executed as planned. A last-minute appeal to the U.S. Supreme Court only delayed the inevitable; Chief Justice Melville Fuller ruled that the New York electric execution law did not violate the Constitution and should stand. William Kemmler would be the first human being to be executed with electricity.

The execution was scheduled for sometime between August 3 and August 6, 1890, the precise time kept secret until hours before the sentence was carried out. When the call went out for the state's official witnesses to report to the prison on August 5, crowds began to assemble outside the prison gates. Kemmler was informed that he would be executed the following morning at 6:00 A.M.

Kemmler was taken out of his cell before dawn and was led to the death chair, the fruit of Harold Brown's dark labors. "Gentlemen," Kemmler said, "I wish you all good luck. I believe I am going to a good place, and that I am ready to go." Kemmler finished his speech with a bow and was placed into the chair. "Now take your time and do it all right, Warden," Kemmler said. "There is no rush. I don't want to take any chances on this thing, you know."

A headpiece was affixed to Kemmler's skull, which made the contraption look like a medieval torture device. Leather bands were wrapped around Kemmler's forehead and chin, partially concealing his features. Eleven leather straps were tightened around Kemmler's arms, legs, and torso. The connections were checked and rechecked. The moment had come.

"Good-bye, William," the Warden said, which was the signal to a man standing by the power switch. The lever was thrown and

Kemmler's body stiffened as 1,700 volts of alternating current from a Westinghouse dynamo surged through every nerve ending. Kemmler's body was rigid "as though cast in bronze," a *New York Times* reporter wrote, save for the index finger of his right hand, which closed up so tightly that the nail pierced the skin and blood trickled onto the arm of the chair. A doctor stood next to Kemmler holding a stopwatch. Five seconds passed, ten seconds, fifteen. At seventeen seconds, the warden pressed a signal button and the Westinghouse dynamo whirred to a stop. The doctor pronounced Kemmler dead.

The announcement, however, proved to be premature. Kemmler stirred in the seat and let out a low animal groan. "Great God, he is alive!" one witness cried. "Turn on the current!" screamed another. A reporter from one of the press associations, unable to bear the sight, fainted on the spot.

The Westinghouse dynamo was hastily restarted, and Kemmler was subjected to another 1,700-volt burst. This time, the dynamo wasn't running smoothly, and the current crackled as it entered Kemmler's body. Blood began to appear on Kemmler's face like crimson sweat, and smoke rose from the top of his head. The skin and hair beneath the electrodes began to sizzle as the sickening odor of burning flesh filled the room. No one knew exactly how long the second jolt of current was applied—witnesses wearing watches were too horrified to consult them. When the current was finally switched off, William Kemmler's name had been forever burned into the history books as the first person to die in the electric chair.

A reporter tracked down George Westinghouse in Pittsburgh and asked about the execution. "I do not care to talk about it," a shaken Westinghouse said. "It has been a brutal affair. They could have done better with an axe." Then, perhaps sensing he had not defended alternating current enough, Westinghouse added, "The public will lay the blame where it belongs and it will not be on us. I regard the manner of the killing as a complete vindication of all our claims."

Both Edison and Harold Brown maintained that Kemmler had been killed painlessly in the first seconds that the current flowed—

the rest of the procedure was merely applying current to a dead man. Edison suggested, however, that future executions be conducted with even more powerful Westinghouse generators that would be kept running continuously. And he suggested a new name for the procedure. Henceforth, condemned men would be *Westinghoused*.

The name, of course, never caught on. Condemned men were electrocuted, fried, zapped, baked, burned, and made to ride the lightning, but never Westinghoused. By the time New York's second electrocution took place the following spring, the death chair sported a more powerful Westinghouse dynamo and thicker wires. The electrodes were placed on the condemned man's calf rather than at his spine, so the current would pass through the heart, and the dynamo was kept running continuously.

The next several executions went comparatively smoothly, and in a surprisingly short time, the electric chair came to be considered an acceptable and even humane means of carrying out death sentences. Edison's home state of Ohio introduced electrocution in 1896, followed by Massachusetts in 1898 and Edison's adopted state of New Jersey in 1906. Soon, more than twenty states were using electric chairs.

Old Sparky, people called it. New York State would go on to use the electric chair for seventy-two years, eventually sending 695 people to their deaths. The executioners settled on a formula for the condemned: 2,000–2,200 volts of alternating current at 7–12 amperes for about twenty seconds, lowered and reapplied at various intervals until death. But prison officials learned what Harold Brown already had discovered, that electricity was very unpredictable, to say nothing of the people charged with administering it. In 1946, convicted murderer Willie Francis was severely shocked but not killed by the Louisiana electric chair, reportedly shrieking "Stop it! Let me breathe!" as the current was applied. It turned out that an intoxicated guard had improperly wired the chair. After an unsuccessful appeal to the U.S. Supreme Court, Francis was returned to the chair a second time and executed.

The horrors continued. The May 4, 1990, electrocution of murderer Jesse Tafero was marked by an unexpected power surge that caused a six-inch long tongue of flame to shoot from the condemned man's head. Alabama killer Horace Dunkins was burned to death before the electric shock could kill him after the cables were connected to the wrong wall receptacles.

Positive and negative. For the electric chair, the flow began to reverse in the late 1970s. After a string of botched electrocutions, the one-time scientific wonder seemed a barbarous relic; it was the same argument that had retired the hangman's noose a century before. In 1982, Texas abandoned the electric chair in favor of lethal injection, and many states soon followed suit.

Currently, only eight places on the planet still use electricity to kill criminals, all in the United States: Alabama, Arkansas, Florida, Kentucky, Nebraska, South Carolina, Tennessee, and Virginia. In Nebraska, electrocution remains the only method of execution; inmates in the other states are given a choice between the electric chair and lethal injection. So far, all but one inmate given the choice have opted for lethal injection.

9

PULSE OF THE WORLD

George Westinghouse needed some good news. The AC standard he had worked so hard to establish now carried the stench of death. Electricity always had the power to kill—that was fundamental to its dual nature. But Harold Brown had managed to portray AC as something that could only take life. Westinghouse searched for a way to show the public the other half of the story.

He found an answer thousands of miles away in the tiny Colorado mining town of Telluride. Once a leading center for gold mining in the Rockies, Telluride had fallen on hard times, squeezed by spiraling energy costs. The mines used heavy excavating machinery that consumed enormous amounts of power, and the mining companies had already used up cheap sources of fuel. There was still gold in the hills, but the extraction costs were becoming too expensive to mine it.

If the mines closed down, Telluride would disappear with it. No one understood this better than Lucien Nunn, the owner of the local San Miguel County Bank. Nunn was an Ohio native who'd attended Oberlin and Harvard before traveling west to seek his fortune. Barely five feet tall, Nunn had hummingbird-like energy; at various times in his life, he built cabins, ran a restaurant, practiced law, and published a local newspaper.

Nunn knew that the runaway cost of energy at the Telluride mines posed a grave threat to the entire town. He had followed the growth of electricity on the East Coast with great interest, and wondered whether new technology might solve Telluride's energy problems. Three miles away from the mines, the San Miguel River roared down a mountainside, a potential source of cheap and abundant power.

The Gold King Mine appointed Nunn, along with his brother Paul, to design a power plant by the river. Nunn knew that a direct current plant was out of the question; DC couldn't be transmitted the three miles from the river to the mines. Nunn took a chance that Westinghouse's alternating current system was the answer. Nunn personally went before the Westinghouse board in Pittsburgh to request the necessary equipment to build an AC power plant and long-distance transmission lines.

George Westinghouse was happy to oblige; Nunn's project could serve as a powerful proof of his AC concept. If an AC system could send power over a remote section of the Rockies, it could do it almost anywhere. In the summer of 1890, Westinghouse sent a 3,000-volt AC generator and a 100-horsepower Tesla motor to Nunn at Telluride. Nunn hired students from Cornell University to help build what eventually became the Ames power plant.

On June 19, 1891, water from the San Miguel River was unleashed onto a six-foot-tall water wheel. The wheel was attached by a belt to a Westinghouse generator whose armature began to rotate as the wheel turned. The alternating current produced by the rotating armature was transmitted three miles to power a mill used to crush rock at the Gold King Mine.

The Ames plant literally saved the town of Telluride. More important, it was proof that the AC system had arrived. The economy of generating power in one location and transmitting it to where it was needed was clear for all to see. The Ames plant turned out to be not only economical but also incredibly reliable. The original plant, much upgraded over the years, is still producing power today.

Electrical companies worldwide began to take notice of AC's proven capabilities. In August 1891, a 30,000-volt polyphase AC system transmitted electricity from Lauffen, Germany, to the site of an international electrical exposition in Frankfurt. The transmission distance was a staggering 106 miles, easily the longest AC line ever. The electricity produced at Lauffen illuminated a display of a thousand incandescent lamps in Frankfurt, a day's journey away. Viewers could only gaze in wonder. Never had electricity traveled so far in the service of mankind.

It was the dawn of a new era; power freed from geography. The Industrial Revolution had so far taken root in cities and towns that were close to sources of energy such as coal, wood, and hydropower. Now areas located far from power sources could become industrial centers. It was as though, as one engineer of the day put it, every town now stood "on an inexhaustible field of smokeless, dustless coal." The Lauffen transmission convinced the city of Frankfurt to adopt alternating current for its municipal power plant, and other European cities followed suit.

The advances being made by AC did not go unnoticed by Edison, or his company. Henry Villard, president of the Edison General Electric Company, wrote to Edison about the AC breakthroughs in Europe, hinting strongly that the company should at least look into developing an alternating current system as a complement to its DC offerings. Edison scowled and wrote back, "The use of alternating current instead of direct current is unworthy of practical men." Practical men, however, were just the sort of people who saw AC's potential.

And Edison's opinions didn't carry the weight they once did. The Wizard's days as the master of the company that bore his name were numbered. Edison's lamp works in Newark, New Jersey, and his machine shop in Schenectady, New York, were consolidated in 1889 to form the Edison General Electric Company. Although Edison was the public face of the company, he owned only about 10 percent of the firm's stock. The rest was controlled by Wall Street bankers, among them J.P. Morgan. Henry Villard was a financier himself; he had organized the highly profitable Northern Pacific Railroad, and like Westinghouse, was more a dealmaker.

Villard and the moneymen behind the Edison General Electric Company had grown increasingly frustrated with Edison's refusal even to consider alternating current. So Villard went behind the Old Man's back and opened up secret merger negotiations with

Thomson-Houston, a rival electrical firm that had substantial investments in AC technology.

The merger talks were also prodded by a cash crunch at Edison General Electric. Many fledgling power companies had purchased Edison equipment but hadn't yet turned a profit. The Edison company had healthy revenues approaching $11 million and still held valuable patents, but it was stretched too thin. The firm tried to make cuts up and down the organization, some of them ruthless. On a wintry morning in January 1892, 150 young women working in the fiber shop of an Edison factory in Harrison, New Jersey, were greeted by a terse note as they knocked off work: "This department is closed." The women, who earned $8 to $10 a week, were replaced by a group of Polish immigrants at half the wage.

Edison was never motivated by money, but the prospect of his company skidding into insolvency unnerved him. Casting a weary eye over the company's balance sheet, Edison commented to colleague Samuel Insull, "This looks pretty bad. I think I could go back and earn my living as a telegraph operator." Behind the nervous joking was the fear of failure. When word of the merger talks with Thomson-Houston reached him, Edison wrote an unusually impassioned letter to Villard.

"If you make the coalition [with Thomson-Houston], my usefulness as an inventor is gone," Edison wrote. "My services wouldn't be worth a penny. I can only invent under powerful incentives. No competition means no invention. It's the same with the men I have around me. It's not the money they want but a chance for their ambition to grow."

At the same time, Edison was tired of big business and the constant patent battles waged over his devices. He preferred the unstructured life of an inventor, free to pursue whatever area of inquiry fired his imagination. Increasingly, he had felt like a small cog in the vast machine of his own creation. In 1890, Edison wrote to Villard: "I feel that it is about time to retire from the light business and devote myself to things more pleasant, where the strain and worry is not so great." There were other projects Edison was keen to tackle—

making improvements to his phonograph and developing his talking-picture kinescope, along with an ambitious plan for magnetic ore separation.

Edison also seemed to sense that the world of electricity was passing him by. A door was closing in the electricity market, and he stubbornly refused to step through it. His resistance to AC now seemed to be little more than stubborn inflexibility. He was no longer a leader in the field of electricity but rather a cranky old-timer resistant to change, the sort of figure that Edison had always hated as a young experimenter. One day in 1890, when a lab assistant asked Edison a question about electricity, the inventor replied that it would be better to consult his chief electrician, Arthur E. Kennelly. "He knows far more about electricity than I do," Edison said with surprising bitterness. "In fact, I've come to the conclusion that I never did know anything about it."

Edison's Wall Street backers pressed harder for a merger with Thomson-Houston, driven by the realization that the electricity business was becoming a regional monopoly controlled by a handful of large players. Most cities already had a single gas company and telephone firm; financiers liked the idea that the electricity market was organizing along similarly profitable lines. By combining several companies into a large electrical trust, the conglomerate pooled the patents owned by each company, putting it in a commanding position.

On April 15, 1892, the deal was struck: Edison General Electric and Thomson-Houston combined to form a new company, known as the General Electric Company. The Edison name had been stripped from the company entirely, which stung the inventor badly. Edison learned of the company's new name from his secretary, Alfred Tate, who recalled, "I had never seen him change color. His complexion was naturally pale, but following my announcement it turned as white as my collar."

At the time of the deal, the Thomson-Houston companies were valued at $17 million, while Edison General Electric Company was worth about $15 million. It was more a takeover than a merger, and

Thomson-Houston executives dominated the new company. General Electric's first president was Charles Coffin, the former head of Thomson-Houston, a one-time shoe salesman. Edison was given a token seat on the company's board of directors.

"Something had died in Edison's heart," said Tate. "He had a deep-seated, enduring pride in his name. And this name had been violated, torn from the title of the great industry created by his genius through the years of planning and unremitting toil."

The creation of GE and the removal of Edison's name from the company he founded left the inventor despondent for a time. But Edison was never one to brood over setbacks. Many of his greatest inventions sprang from what seemed like utter defeat. "Edison seemed pleased when he used to run up against a serious difficulty," one of his employees recalled. "If it fails on its merits, he doesn't worry or fret about it, but, on the contrary, regards it as a useful fact learned; remains cheerful and tries something else. I have known him to reverse an unsuccessful experiment and come out all right."

Edison liked nothing better than to snap his suspenders against his chest and match wits with an opponent. He relished a good fight, especially one in which the battle lines were clearly drawn. The General Electric Company no longer bore Edison's name, but there was plenty of his sweat still tied up in the company. Before long, Edison was once again on the prowl for a chance to prove his detractors wrong and hand Westinghouse a stinging defeat.

The opportunity wasn't long in coming. The city of Chicago had announced plans to hold a grand fair to commemorate the four hundredth anniversary of America's discovery by Christopher Columbus. The Columbian Exposition was to be the largest of its kind ever held in America, and electricity was to be the star attraction. Chicago fair organizers planned to design buildings around the artful use of artificial illumination, and to use electricity as the fair's exclusive power source.

The contract to provide power and light to the fair was put up for bid, and the competition was intense. International fairs had emerged as influential shapers of public opinion in the late nineteenth century, the corporate image ads of their day. The fairs were a place for manufacturers to display their wares, make contacts, and most of all, get the public excited about the technology that would make their lives easier. Being chosen as the company to bring electricity to the gaudy affair would be a major coup for the winning concern.

General Electric and Westinghouse immediately locked horns in the bidding war for the fair. Many assumed GE would win the contract because of the company's association with Edison, its long history with incandescent lighting, and its strong presence in the Midwest. But George Westinghouse wanted the contract badly. The prestige that flowed from powering such a high-profile exposition would be priceless, especially for a company still trying to make a name for itself. Westinghouse submitted a low-ball bid that undercut General Electric's offer by more than half, and in May 1892, was awarded the contract to provide power and light to the fair.

Edison was miffed at losing the bid, and launched a petty rearguard action aimed at crippling Westinghouse's efforts at the fair. Edison brought suit against the Westinghouse company, claiming it was infringing on several of Edison's long-standing patents that covered incandescent light bulb design. A crucial design element of the incandescent lights Westinghouse planned to use to illuminate the fair was that the bulb was made in one piece, with the glass bottom fused to the wires, preserving a near-perfect vacuum. The courts ruled that Edison held the exclusive rights to the one-piece design, and ordered Westinghouse to stop making bulbs fashioned in that manner.

The ruling couldn't have come at a worse time for Westinghouse. With the exposition less than a year away, he didn't have a single light bulb he could legally use at the fair. George Westinghouse, however, was at his best when confronted with big problems; he was a man given to oversized solutions. Westinghouse took the best bulb patent he owned—a two-piece design known as the Sawyer-Man lamp—and put his entire company on a crash course to come up with a modified design of the bulb to use at the fair.

Westinghouse engineers came up with a way to seal the globe using a glass stopper bottom that would hold a vacuum and keep the filament from burning up, while still skirting Edison's patents. The two-piece design wasn't nearly as good as the Edison bulb, but it might be just good enough. Westinghouse built a new glass factory for the project and churned out a quarter of a million lamps in less than a year, a remarkable marshaling of manufacturing resources.

When the fair opened on May 1, 1893, visitors entered an electrical wonderland the likes of which they had never seen. One hundred thousand people jammed the Court of Honor to watch President Grover Cleveland turn a golden lever that sent the Westinghouse dynamo engines into motion, powering the fair's hundreds of thousands of lamps and all of its machinery. The spectacular lighting bathed the fairgrounds in a magical glow; children's author L. Frank Baum was so enthralled by the sight that he used it as inspiration for the Emerald City in his *Wizard of Oz* book series.

The dazzling display seemed to point the way to a brighter future. "Among monuments marking the progress of civilization throughout the ages, the World's Columbian Exposition of 1893 will ever stand conspicuous," solemnly intoned *The Book of the Fair*. "Gathered here are the forces which move humanity and make history, the ever-shifting powers that fit new thoughts to new conditions, and shape the destinies of mankind."

About 27 million people visited the Exposition, nearly a quarter of the country's population at the time. The Ferris Wheel made its debut; for 50 cents, riders were packed sixty to a car and hoisted 264 feet in the air, giving them a commanding view of the fair's buildings and outdoor fountains bathed in brightly colored searchlights. Several soon-to-be-famous consumer products were introduced at the fair: Aunt Jemima Syrup, Cracker Jacks, Shredded Wheat, and Juicy Fruit gum.

The exhibitions that made the most vivid impression on visitors were housed in the Electricity Building. The hall was a cathedral of electricity; visitors entered the building by walking past a frieze bearing the inscription *Eripuit Coelo Fulmen Sceptrumque Tyrannis:*

"He snatched lightning from the sky and the scepter from the tyrant." It was a phrase once used to describe America's founding father of electricity, Benjamin Franklin. Inside the hall, the electrical displays were designed to erase the fear many had of the technology and replace it with a sense of wonder. There was an electric moving sidewalk, an elevated electric train, an electric kitchen, and tens of thousands of incandescent lights. (The electric moving sidewalk was often out of service, giving fair goers a more balanced view of the electrical future.)

Inside the Electricity Building, Westinghouse and General Electric squared off eye to eye; the companies' respective displays were adjacent to each other in the same hall. George Westinghouse had put considerable care into showing the public just how far his company had come in less than a decade in business. A large display proclaimed "TESLA POLYPHASE SYSTEM," giving the relatively unknown Tesla equal billing with the nationally famous Westinghouse.

Westinghouse displayed a complete polyphase electrical system at the fair. There was an AC generator with transformers for raising the voltage for long-distance transmission, a short transmission line, another set of transformers to step down the voltage, and a rotary converter that transformed some of the AC power into direct current for engines that still ran on DC, such as railway motors. This, in miniature, was the system with which Westinghouse intended to power the world.

Tesla had a display of his own, including an unusual exhibit dubbed the "Egg of Columbus," used to explain the principle of the rotating magnetic field and the induction motor. The device consisted of a series of polyphase electrical coils hidden beneath a plate on which rested a copper egg. When the coils were energized, a rotating magnetic field was created, causing the egg to stand up on its end. Tesla also displayed the first neon light tubes at the exposition, which used high-frequency currents to bring gases inside the glass to incandescence. Tesla had the tubes hand-blown to spell out "WELCOME ELECTRICIANS" in glowing letters.

Adjacent to the Westinghouse display stood the General Electric Company exhibit. It was a massive collection of electrical equipment set off by the imposing Edison Electric Tower, a tall white shaft encircled by thousands of miniature lamps, which reflected light off of shards of crystal. Next to the tower was a display showing off 2,500 specimens of Edison incandescent lamps, the same bulbs that Westinghouse had been prevented from using to illuminate the fair.

The basic design of the Edison lamps, however, was already more than a decade old. For the most part, the General Electric display only served to demonstrate how much the company was trading on past glories and how far the Westinghouse company had moved ahead of it. To Edison's dismay, one of General Electric's exhibits featured a polyphase AC system on display. Alternating current from an Edison company—it was only possible now that Edison had practically no control over the company he founded.

The most impressive exhibit of all, the one that really changed the course of technology, was a display that practically no one saw: the massive Westinghouse machinery that powered the entire fair, hidden in the bowels of the Hall of Machinery. The Westinghouse generating plant for the fair was the largest AC central station then in existence, and the first large polyphase system ever built in the United States. It was the first truly universal AC system, able to power incandescent lights, arc lamps, and other DC applications through use of a rotary converter. Everything that moved or lit up at the fair was powered by the Westinghouse polyphase AC system—even General Electric's exhibit.

The fair would prove to be an important victory for Westinghouse and a turning point in the public's perception of alternating current. In the year following the fair, more than half of all new electrical devices ordered in the United States ran on alternating current, largely due to Westinghouse's success and the superior performance of Tesla's induction motor at the exhibition.

Tesla himself attended the Chicago fair, giving a lecture before an audience of electrical engineers and scientists. Tesla also performed public demonstrations at the fair, astonishing crowds by taking 200,000 volts of alternating current into his bare hands, enveloping his body in a dazzling stream of light. The electricity didn't hurt Tesla because the extremely high-frequency current he produced traveled along the surface of his skin, rather than through his body. Negative and positive: Harold Brown had taken 200 volts of AC and used it to kill. Tesla handled a thousand times more voltage in order to educate and amuse.

The Chicago fair put the Westinghouse forces in a leading position for an even bigger project, one that Tesla had dreamed about since he was a boy: Niagara Falls. The falls had long been an inviting source for generating power. About one-fifth of the U.S. population lived within four hundred miles of Niagara, and Buffalo (a city of 250,000) was only twenty miles away. The flow of water over the falls was steady and reliable, making it ideal for spinning a turbine smoothly to produce a continuous flow of electricity.

Designing a power plant at the falls that could transmit electricity many miles away posed enormous technical challenges, and local officials at first turned to the country's most famous electrical expert, Thomas Edison. In November 1889, while Harold Brown was feverishly conducting his animal experiments at Edison's lab, the inventor submitted a plan for building a DC power station and distribution system at the falls. Westinghouse declined to submit a plan at the time, saying he doubted whether electrical power—AC or DC—could be transmitted to Buffalo cheaply enough to compete with the steam power then widely in use.

To evaluate the proposals, a five-man International Niagara Commission was appointed, headed by one of the leading physicists of the day, Sir William Thomson, soon to be Lord Kelvin. Thomson was a DC man through and through. He had grown up with direct current and considered AC an unproven and unnecessary alternative. The commission invited twenty-eight firms in the United States and Europe to submit plans to harness the falls, and offered a $3,000 prize for the winner.

Neither Edison nor Westinghouse submitted a formal proposal to the commission. Edison had been suggested as a commission member, and possibly thought that the commission would eventually have to turn to him for help. Westinghouse's engineers urged their boss to enter the contest, but Westinghouse was unwilling to reveal the company's trade secrets for AC transmission with no guarantee of a deal. "These people are trying to get $100,000 worth of information for a prize of $3,000," Westinghouse declared. "When they are ready to do business, we will submit a plan and bid for the work."

As it turned out, the commission ruled that all the submissions fell short of offering a complete plan for both power production and distribution at Niagara. With the prodding of Lord Kelvin, the Commission voiced its doubts about AC, reporting that they "were not convinced of the advisability of departing from the older and better understood methods of continuous currents in favor of the adoption of methods of alternating currents."

The commissioners, though, represented electricity's old guard; the marketplace had already begun to embrace AC. Westinghouse's successful power plant at Telluride and his triumphant powering of the Columbian Exposition convinced many that alternating current wasn't the risky, unproven technology that men like Edison and Lord Kelvin thought it was. Kelvin, like Edison, stubbornly clung to his preference for DC to the end, ignoring the evidence in front of his nose. "TRUST YOU AVOID THE GIGANTIC MISTAKE OF ALTERNATING CURRENT" Kelvin cabled commissioners from England in May 1893. But backers of the Niagara project rejected the advice of the great physicist and the world's greatest inventor and came out in favor of an AC system for the falls. It was a triumph not just for the Westinghouse company but for alternating current as a technical standard.

Westinghouse wound up providing the AC generators, switchgear, and auxiliary equipment for the power plant at Niagara. General Electric was given a decidedly secondary role in the project, supplying transformers and maintaining the AC transmission line to Buffalo. The first two Westinghouse generators roared into service in August 1895, sending alternating current crackling down the line to Buffalo and beyond.

For Tesla, the Niagara project was the culmination of a life-long dream; as a youth he had vowed to travel to far-away America and harness the energy of Niagara Falls. "Thirty years later, I saw my ideas carried out at Niagara and marveled at the unfathomable mystery of the mind," Tesla said. Few projects in Tesla's wide-ranging career gave him more satisfaction; he would later liken the Niagara project to the building of the pyramids, "a monument worthy of our scientific age, a true monument of enlightenment and of peace."

Once the Niagara plant was in operation, delivering electricity to nearby Buffalo turned out to be less important than everyone thought. The availability of cheap and abundant power spurred industrial development throughout western New York, and before long, power was being sent to New York City, more than 450 miles away. In the decades to follow, electricity from the falls transformed Detroit into the Motor City, powering the city's assembly lines and steel furnaces. Niagara power spawned an entirely new industry, the electrochemical business, which used massive amounts of electricity to produce caustic chemical compounds such as chlorine. The Union Carbide Company was for many years one of the Niagara plant's biggest customers. Today, the Niagara plant on the American side of the border, much expanded, is still producing electricity.

Niagara became the model for the way electrical power would be generated and consumed in the twentieth century and beyond. Electricity would be produced wherever there was a source of reliable power, transmitted hundreds, even thousands of miles, and consumed where it was most needed. Niagara removed the last serious doubt about the efficiency of the AC system. After Niagara, even more ambitious hydroelectric plants were built at the Hoover and Grand Coulee Dams, soaring concrete and steel monuments to

America's ingenuity and growing world power. Both power stations generated electricity using Westinghouse equipment.

George Westinghouse was no genius; the label better fit Edison and, in his own way, Tesla. But Westinghouse was smart enough to know the limits of his own intelligence. Surrounded by genius, George Westinghouse listened and learned. When new answers revealed themselves, he was ready for them. George Westinghouse had sensed the pulse of the future, and it was alternating.

10

KILLING AN ELEPHANT

Edison's DC standard was rapidly slipping into irrelevancy, but the Old Man still had some fight left in him. Even though General Electric had become just another firm hawking the "deadly" alternating current, Edison longed to fight another round. Seeing DC all but lose the war of the currents was a rare and unsettling defeat for the Wizard, and Edison kept a sharp eye out for an opportunity to stick it to his AC opponents.

Early in 1903, he got his chance. The situation was appropriately circus-like: Edison agreed to lend his technical expertise to the public electrocution of a rogue Coney Island elephant named Topsy. It would be Edison's final public demonstration of the killing power of alternating current and the most cruelly ambitious; the six-ton elephant was easily the largest creature Edison would attempt to put down with electricity.

The projected victim was an ill-tempered circus elephant that had been brought to America in 1885 as part of the Adam Forepaugh Circus, a rival to Ringling Brothers. Topsy, eight years old when she first came to the United States, was exhibited as "The Original Baby Elephant" in a traveling circus that made grueling tours of small towns across the country. After several years, Topsy could no longer play the role of baby elephant convincingly—she stood ten feet high, was nearly twenty feet long, and the circumference of her leg alone was two feet. Topsy was recast as a performing elephant, taught a variety of tricks by a succession of handlers.

In 1900, after fifteen mostly uneventful years under the big top, Topsy became unmanageable. During a scorching summer tour through Texas, Topsy unexpectedly turned on her trainer and stomped him to death. A new handler was hired, but Topsy didn't care much for him, either. During a show in Paris, Texas, Topsy crushed the new trainer to death with her leg. Topsy, worth more than $6,000, was too valuable to let go. Elephant handlers were replaceable; six-ton trained circus elephants were not. Even after killing two men, Topsy continued to tour with the circus, although her new keepers now kept a wary distance.

Two years later, when the circus rolled into Brooklyn, New York, Topsy's latest keeper, J.F. Blount, came up with the ill-conceived idea of feeding a lighted cigarette to Topsy as part of a planned act. Topsy reacted by lifting Blount in the air with her trunk and slamming him to the ground, killing him instantly.

Having killed three handlers in as many years, Topsy was sold to Luna Park, a Coney Island development then under construction. Her new handler was Whitey Alt, a man of whom newspapers would later say "had a habit of taking more stimulant than was good for him." Pairing a killer elephant with a lush trainer was a combustible mixture. One evening, a tipsy Alt led Topsy on an impromptu walking tour of Coney Island, winding up at the police station, where Topsy got stuck trying to cram her head through the front door. Alt was relieved of his duties, and Topsy's days as a performing elephant were numbered.

While Coney Island officials debated what to do with the troublesome elephant, Topsy was put to work lifting heavy wooden beams used in the construction of Luna Park, an ambitious development that would feature a scenic railroad, a carousel, and live animals including elephants and ostriches. Finally, Topsy's owners found a Manhattan man willing to buy the elephant's hide, tusks, and other body parts provided the park killed the animal. An execution date was set in January 1903, and a large wooden scaffold was constructed for the improbable purpose of hanging Topsy by the neck.

Before the hanging could take place, the SPCA intervened, just as it had done fifteen years before when Harold Brown electrocuted a dog in public at Columbia College. The organization argued that hanging a six-ton elephant was an absurd proposition and quite likely to be botched, resulting in needless suffering. It was the same argument that had been raised more than a decade before about the hanging of criminals.

Not by coincidence, it resulted in the same solution. When Luna Park officials put out the word that Topsy would be killed by more humane means—electricity—Thomas Edison quickly offered his services. Edison dispatched three of his top electricians to serve as Topsy's executioners. The electricity used to kill the elephant—alternating current, of course—would be supplied by Coney Island's own generator that provided power and light to the amusements.

The execution was set for the afternoon of January 4, 1903, five months before Luna Park officially opened for business. Shortly before 1:30 P.M., Topsy was led to the scaffold originally built for her hanging, which had been converted into a makeshift electrocution platform. Two wires stretched from the scaffold to a nearby building where the AC generator was housed; at the end of the wires were two large electrodes designed specially for the occasion. When Topsy reached the narrow approach to the scaffold, however, she balked and refused to walk any farther. Her former handler, Whitey Alt, was summoned and offered $25 if he would help coax the elephant across the narrow approach. Alt refused, saying he wouldn't do it for twice that amount. The scaffold was abandoned and the electrocution site was moved to a nearby courtyard.

By the time Topsy was moved into position, the execution was already running more than an hour behind schedule. The three Edison electricians struggled to affix the electrodes to Topsy's legs, ducking under her body as several other men held the elephant in place with ropes. The electricians finally got Topsy to lift her leg—one of her circus tricks—so that the connections could be made, one electrode attached to the right front leg, the other to the left hind leg. Copper-clad sandals were secured to her feet to serve as electrical conductors. At 2:38, a veterinarian fed Topsy two carrots laced with 460 grains of cyanide, which the elephant greedily wolfed down.

The cyanide wasn't enough to kill Topsy; 6,000 volts of alternating current would have to finish the job.

The three Edison electricians waited for the signal to throw the switch. As they did, a fourth Edison employee scurried into position. In his hands was one of Edison's latest inventions, a device that would come to prominence in the decades to come: the motion picture camera.

Motion picture technology was still so new at the time that few observers understood what the Edison man was doing when he peered through his strange-looking device and slowly began turning a hand crank. The cameraman went largely unnoticed by the crowd of fifteen hundred onlookers that had gathered to witness Topsy's killing; indeed, the accounts in the next day's newspapers made no mention of the cameraman at all. As the crank on the camera turned, the event was instantly transformed. A dreary episode witnessed by a small throng of curiosity-seekers suddenly became a moment forever preserved in time.

The motion picture camera was a typical Edison invention; it drew on previously developed ideas, but it advanced them in such a novel fashion that it represented a true breakthrough. In 1887, the notion first occurred to Edison "that it was possible to devise an instrument which should do for the eye what the phonograph does for the ear, and that by a combination of the two, all motion and sound could be recorded and reproduced simultaneously." Edison's goal of a moving picture with a synchronized soundtrack wouldn't be realized commercially until the 1927 release of *The Jazz Singer*, but it would take the inventor little more than three years to achieve what no one else had: a camera that recorded continuous motion on film.

The closest anyone had come to capturing motion on film was the work of English-born photographer Eadweard Muybridge, who began a pioneering study of animals in motion in 1872, sponsored by wealthy California railroad baron Leland Stanford. Stanford, a one-time California governor, was a racehorse enthusiast who had a pet theory that a horse in full gallop would, at one point, have all four hooves completely off the ground. There was no way to test Stanford's notion with a conventional camera—the horse's motion was simply too fast for any still camera to capture. So Stanford commissioned Muybridge, a respected California landscape photographer, to come up with a way to capture a horse's full range of motion.

On June 15, 1878, Muybridge unveiled his solution, an unwieldy multi-camera apparatus that he installed at Stanford's horse track in Palo Alto, California, today the site of Stanford University. Muybridge arranged twelve still cameras in a row alongside the track and attached the shutters of each camera to trip wires stretched across the track. A horse racing down the track would trip each wire in succession and create a sequential photographic record of itself in motion.

As a crowd of horse enthusiasts and journalists looked on, one of Stanford's prize horses galloped down the track, and Muybridge's cameras captured three complete strides of the animal. When the photos were developed, it proved that Stanford's theory was right on the money: all four of the horse's hooves left the ground in mid-gallop.

The Muybridge apparatus was clever, but its applications were limited. To produce a motion picture lasting just one minute would have required using 720 cameras. Edison recognized Muybridge's device as a technological dead end. The inventor wanted to build a single unit that would do the job of Muybridge's multiple cameras.

In 1888, Edison set out to invent the motion picture camera using his preferred method of inquiry—exhaustive trial and error. "We tried various kinds of mechanisms and various kinds of materials and chemicals for our negatives," Edison later recalled. "The experiments of a laboratory consist mostly in finding that something won't work. The worst of it is you never know beforehand, and sometimes it takes months, even years, before you discover you have been on the wrong line all the time."

Initially, Edison thought a motion picture camera could be built along the lines of his phonograph. He designed a cylindrical disk that had a photosensitive coating and tried embedding microscopic photographs on the disk that would be enlarged on playback. But the coatings that Edison tried, including dry albumen and silver emulsion, produced images that were too coarse to withstand intense magnification.

Edison soon abandoned the disk altogether in favor of a new material that had recently come on the market: celluloid. It was a natural resin drawn from plant fibers that produced a film that was flexible and surprisingly resilient, although quite flammable. Edison took a narrow sheet of celluloid and imprinted it with a series of photographs arranged spirally, and then stretched the celluloid over a cylindrical drum. When the drum was turned, the images flashed by sequentially. The cylinder had its own limitations. The photos imprinted on the celluloid were very small, and only the center of each image could be properly brought into focus.

Edison rarely sought outside help with his inventions, but the motion picture camera was particularly complicated, requiring new advances in both the mechanical and photographic arts. For help with the science of photography, Edison turned to George Eastman. The man who would soon found the Eastman Kodak Company was already a recognized leader in photography. In 1884, Eastman patented the first practical film in the form of a roll; four years later, he produced the first Kodak camera, specifically designed for his roll film.

Eastman was working on a new type of dry film that Edison thought had promise as film stock for his movie camera. Eastman custom-made a narrow strip of fine grain film for Edison, saving the inventor countless months scouring the globe for the right chemical combination. "Without George Eastman, I don't know what the result would have been in the history of the motion picture," Edison would later state, rare praise from a man who seldom shared credit for any of his inventions.

With the film problem on the way to being solved, Edison turned to the equally important challenge of designing a mechanism that could advance the film through the camera with split-second precision. The instant a single frame was exposed, the camera would have to move the film into position to expose the next frame—all in about 1/100th of a second.

"This had to be done with the exactness of a watch movement," Edison recalled. "If there was the slightest variation in the movement of the film, or if it slipped at any time by so much as a hair's breadth, this fact was certain to show up in the enlargements."

For aid with the mechanical aspects, Edison leaned heavily on one of his laboratory assistants, W.K.L. Dickson, who had a background as a photographer and would later become one of Edison's chief cameramen. Dickson and Edison experimented with different rates of speed for the film as well as various film sizes. As usual, Edison's early work would set the standard for an entire industry. Initially, he designed a camera that ran at forty-six exposures per second, but soon decided that the most efficient speed was in the range of twenty to twenty-five frames per second. (Modern theatrical films run at twenty-four frames per second.) Edison's choice for film width was equally influential. After testing numerous film sizes, Edison decided that a strip of celluloid film, leaving room for sprocket holes, should be exactly 35 millimeters wide. It remains the industry standard more than a century later.

By 1889, Edison had completed most of the work on his movie camera, which he called the Kinetograph. But he didn't patent the device until nearly two years later. "I was very much occupied with other matters," Edison later wrote, with characteristic understatement. The "other matters" involved Edison's bitter battle with George Westinghouse, which had entered its most fevered stage. It was the height of Harold Brown's animal experiments and the effort to have AC adopted as the executioner's current in New York. The motion picture camera would have to wait.

When Edison took up the camera again several years later, he added a device called the Kinetoscope, which played the films back. The Kinetoscope was a large wooden cabinet equipped with an electrical motor that moved a fifty-foot band of film through the field of a magnifying glass. Viewers looked at the moving images through a peephole in the top of the cabinet. The Kinetoscope was displayed at the 1893 Columbian Exposition in Chicago, where it turned out to be one of Edison's few triumphs at the otherwise Westinghouse-dominated fair.

There was no market for the machine Edison had created, so he set about inventing that as well. First, he'd need to produce films to display on his Kinetoscope. In February 1893, Edison built the world's first motion picture studio on the grounds of his laboratory in East Orange: a bizarre structure dubbed the "Black Maria." The small building, about twenty-five feet square with a slanted roof, had its foundation set on a pivot so that the entire structure could be swung to follow the course of the sun. The building was covered with tarpaper and the walls inside were painted flat black so that the actors in the foreground were shown in the sharpest possible relief.

"The Black Maria was a ghastly proposition for a stranger daring enough to brave its mysteries—especially when it began to turn like a ship in a gale," Edison remembered. "But we managed to make pictures there. And, after all, that was the real test."

Beginning in 1893, Edison's team churned out a series of short films. Most were well under a minute in length; the first Edison cameras could hold only about a minute and twenty seconds of film. The earliest copyrighted film that survives is *Edison Kinetoscopic Record of a Sneeze, January 7, 1894*, which shows Edison employee Fred Ott sneezing for the camera.

The Black Maria was soon visited by a steady procession of performers. Strongman Eugene Sandow flexed his muscles and struck various poses for the camera in one 1894 short. There were films of a man balancing on a high wire, a woman doing a butterfly dance, a cockfight, and an act from the Buffalo Bill Wild West Show. Another showed boxing champion "Gentleman Jim" Corbett and a sparring partner squaring off in a ring. The match was obviously staged for the camera; throughout the match, Corbett smiles self-consciously at the camera, the first of many unconvincing movie actors to follow. Edison's film crew produced more than seventy-five films in 1894 alone.

W.K.L. Dickson served as cameraman and director for many of the early films shot in the Black Maria. In 1895, Dickson stepped in front of the camera to appear in an experimental short. In the film, he's shown playing a violin into a gramophone horn as a pair of male assistants dance offstage, the first known attempt to produce a motion picture with a recorded synchronized sound track. True sound and picture synchronization would remain elusive, but Dickson and his team were pioneers in this technical quest.

Initially, the Black Maria films were played back on Kinetoscopes, Edison's bulky single-viewer peepshow devices. The first "Kinetoscope Parlor" opened in New York in 1894, featuring five machines lined up in a row. For 25 cents, customers viewed films in each of the five machines.

Edison failed to see motion picture viewing as the group entertainment it would inevitably become. He stubbornly resisted projecting his films on screens or walls because he felt the images weren't nearly as sharp. As with many of his inventions, Edison was much more adept at providing the public with something new than he was at anticipating how people would ultimately decide to use it. Edison wrongly imagined motion picture viewing as a solitary act, when in fact "going to the movies" would soon become an event in itself, quite apart from the content of a film. Movie viewing quickly became a communal event, helping film become the dominant entertainment medium of the twentieth century.

Edison recognized the entertainment value of motion pictures, but he somehow believed that his invention would be put in the service of more lofty endeavors. "I believe that the motion picture is destined to revolutionize our educational system, and that in a few years it will supplant largely, if not entirely, the use of textbooks in our schools," Edison declared.

Against Edison's wishes, motion pictures were soon being shown on screens and walls by film-projecting devices that went by names such as the Mutoscope, the Phantoscope, and the Vitascope, none of which was made by Edison. The Vitascope was the most successful, and its first theatrical exhibition in 1896 at a music hall in

New York City was an overnight sensation. The moving images on the screen were so lifelike that audience members in the front row ducked for cover whenever the action headed their way.

Projection increased audience interest and expanded income; even so, the movie business experienced years of boom and bust. Edison nearly left the business in 1900, pulling back at the last minute on a deal that would have sold his motion picture interests to the rival American Mutoscope and Biograph Company.

Edison was a keen student of the technical aspects of filmmaking, but had little interest in the medium as an art form. Many assumed Edison shot and directed his own films, since they were prominently promoted as being produced by the Edison Manufacturing Company. But Edison was no auteur. His role was closer to what Hollywood would later call executive producer—the man at the top overseeing a team of moviemakers who did the actual filmmaking.

To Edison, the best movie subjects were drawn from real life, and his tastes tended toward the unusual and even freakish. Among Edison's early films were *Boxing Cats*, which depicted two felines wearing tiny boxing gloves squaring off in a small ring, and *The Execution of Mary, Queen of Scots*, which simulated a beheading by use of stop-action photography. A 1901 film, *Electrocution of Czolgosz*, reenacted the execution of the assassin of President William McKinley, showing an actor being strapped into the electric chair at Auburn Prison and then dispatched with a burst of alternating current.

The planned execution of Topsy promised an even more stunning visual spectacle, one that Edison couldn't resist. Topsy's killing was a splendid opportunity to capture powerful images that would not only astonish viewers but also remind them of the killing power of alternating current. On the day of Topsy's execution, Edison's cameraman was given a front-row view of the proceedings. The resulting minute and a half of film, *Electrocuting an Elephant*, would prove to be one of Edison's longest and most arresting motion pictures to date.

As the film opens, Topsy is led through the half-built Luna Park grounds by three handlers, one walking in front and two trailing behind. The procession is deliberate and somber, giving the impression of a condemned man walking the last mile. Topsy has a harness around her head, while her neck and back are draped with thick ropes. The camera slowly pans right to follow Topsy's path. In the background, there are fleeting glimpses of spectators watching the event: a worker sitting atop a construction beam, a small knot of observers craning their heads to see the procession. As the camera continues to pan, it captures a crowd of several hundred spectators dressed in winter clothing standing on a raised wooden platform. Topsy continues her slow walk in the foreground and at one point gets quite close to the camera, her wrinkled face nearly filling the frame.

Then there's a cut in the film, and suddenly Topsy is standing in place on the execution platform, her feet splayed apart. By now, Edison's electricians have affixed the electrodes to Topsy's legs. Two ropes secure Topsy to the ground, and copper sandals further secure her feet. The camera stays on Topsy; in the background a large sign advertises the yet-to-be-opened Luna Park as "THE HEART OF CONEY ISLAND."

Suddenly, Topsy's entire body stiffens and her trunk curls inward. Wisps of white smoke rise from her feet, and quickly form a thick cloud. Topsy tips slowly to her left side, and then crashes to the ground like a felled tree. The cloud of smoke becomes very dense, and for a few seconds, nearly obliterates Topsy. A spectator suddenly cuts in front of the camera and walks quickly out of the frame, unaware of the cameraman. The smoke begins to dissipate, and Topsy is seen lying on her side on the spot where she fell. The camera remains fixed on Topsy. She is motionless, except for her back leg, which twitches several times.

Then, as abruptly as it began, the film was over. It had taken only about ten seconds to kill Topsy, and the electrodes were still warm when her body was dissected on the spot where she fell. Topsy's parts were scattered to the winds—the head was preserved for mounting, the hide sold for leather, the organs donated to a professor of biology at Princeton, the feet used to make umbrella stands. By nightfall, there was nothing left but a few strewn body parts and a dark stain on the ground.

Topsy's electrocution got splashy coverage in the New York newspapers—"BAD ELEPHANT KILLED" was the headline in the New York *Commercial Advertiser*. "A rather inglorious affair," the *New York Times* opined in a front-page story, a phrase that could well have been applied to the entire AC/DC conflict. But the gruesome spectacle failed to have much effect on the public's perception of alternating current. After all, the same alternating current that killed Topsy was also used to power the nickelodeons that later showed the film of Topsy being electrocuted, not to mention all the amusements on Coney Island. Topsy's electrocution turned out to be a passing curiosity, a sepia-toned image that quickly faded from memory. In time, Luna Park would be gone as well, destroyed by a fire in 1944.

One thing would survive, however: the film of Topsy's electrocution. Unlike most short films of the era, Edison's movies were preserved, mainly because they bore the great inventor's name. In 1940, the Museum of Modern Art in New York acquired the surviving nitrate negatives and prints from the Edison Manufacturing Company and undertook a project to copy key titles for public exhibition. In the early 1970s, the original Edison nitrate negatives were transferred to more stable acetate fine grain film, assuring the survival of the collection for generations to come. The Library of Congress also amassed and preserved an extensive collection of Edison films.

Copies of the Topsy film are still kicking around; there's even a DVD box set of Edison's films that includes *Electrocuting an Elephant*. Not far from the spot where Topsy fell, the Coney Island Museum has a copy of the film, which it shows to appalled visitors.

"People are horrified by it, but also kind of fascinated," says Dick Zigun, the tattooed proprietor of the Coney Island Museum. "It's a shocking moment in history."

To watch the film, viewers stand on copper plates, just as Topsy did, and peer through the viewer of a Mutoscope to see the flickering images from a battle long forgotten.

11

TWILIGHT BY BATTERY POWER

By now, it was clear even to Edison's most loyal supporters that DC had all but lost the war of the currents. Direct current plants transmitting power only a mile from the generator couldn't meet America's surging demand for electricity. Only the AC system could send electricity cheaply and efficiently over long distances.

But Edison wouldn't give up on direct current. With the stubbornness that was both his greatest asset and most conspicuous character flaw, Edison still clung to the notion that DC was, in its own way, a superior technology to AC. As a young telegrapher, he had spent long hours with DC batteries; as an adult had built the country's first DC power station. He wasn't about to abandon the standard now. There was something about DC that appealed to Edison's nature; it was straightforward, easy to visualize. If DC wasn't going to be adopted as the universal standard for delivering electricity to homes and businesses, there had to be another important use for it. But what?

For Edison, the answer came chugging into view in the late 1890s: the automobile. The first motorcars, like most new inventions, were built around a set of clashing standards. Some of the early models were powered by steam, others by gasoline or kerosene, still others by electricity in the form of a rechargeable storage battery.

The storage battery—basically a box of direct current—was what immediately attracted Edison's attention. Edison recognized early on that the automobile wouldn't enjoy widespread popularity until the industry settled on a set of technical standards. Beginning in 1899, Edison set his sights on developing a storage battery that would become the worldwide standard for powering automobiles. It was an idea more than a century ahead of its time: the electric car.

The rechargeable storage battery was first developed in 1859 by French chemist Gaston Planté. Unlike a standard primary battery, which converts chemical energy into electricity and eventually exhausts itself, the storage or secondary battery can be recharged once its electrochemical power has been expended. By applying a direct current to the battery's terminals, the electrodes can be returned to their original chemical state, able to supply power again. The early storage batteries were filled with lead and acid that combined to spark a chemical reaction that produced a flow of direct current. The lead-acid batteries were heavy and dangerously corrosive, characteristics that made them particularly dicey for use in an automobile. Edison was convinced he could come up with a better alternative—a lightweight, noncorrosive battery that could power a car for hundreds of miles on a single charge.

"I don't think Nature would be so unkind as to withhold the secret of a *good* storage battery if a real earnest hunt for it is made," Edison declared. "I'm going to hunt."

Edison plunged into his battery work with characteristic gusto. He was in his element; the battery hunt combined his favorite branch of science—chemistry—with his preferred trial-and-error method of investigation. If he could come up with a superior storage battery, it would put direct current back on the map. The storage battery, Edison predicted, would "open up a new epoch in electricity," one in which DC took its rightful place alongside AC as an essential worldwide standard.

The challenge was similar to Edison's hunt for the right incandescent bulb filament. In the case of the storage battery, Edison needed to find the precise chemical compounds that would combine to form a powerful noncorrosive battery. In storage batteries, two metal rods, called electrodes, are connected by wires and immersed in a liquid known as an electrolyte. The metal rods react with the electrolyte to produce a flow of electrons through the circuit. The lead-acid batteries of the day had electrodes made of lead and lead dioxide and an electrolyte consisting of acid. Edison needed to find another set of compounds to replace the lead and acid that wouldn't have the same disadvantages but would still produce a strong current of electricity. The hunt was on.

Walter Mallory, a close associate and vice president of the Edison Storage Battery Company, recalled Edison's dogged search for the ideal battery compounds: "I found Edison at a bench on which there were hundreds of little test cells that had been made up by his corps of chemists and experimenters. He was seated at this bench testing, figuring, and planning. I then learned that he had thus made over nine thousand experiments in trying to devise this new type of storage battery, but had not produced a single thing that promised to solve the question. In view of this immense amount of thought and labor, my sympathy got the better of my judgment, and I said: 'Isn't it a shame that with the tremendous amount of work you have done you haven't been able to get any results?' Edison turned on me like a flash, and with a smile replied: 'Results! Why, man, I have gotten a lot of results! I know several thousand things that won't work!' "

Thousands of compounds were tested, and the results were painstakingly logged in laboratory notebooks. Edison also put his prototype batteries through unusual endurance tests, having his workers toss batteries from second and third story windows to see if they could withstand the fall without leaking. After more than three years of experiments, Edison finally settled on a winning combination: nickel as the positive electrode, iron as the negative electrode, and an alkaline solution of potassium hydroxide as the electrolyte.

The new "E" battery hit the market in 1904 and Edison wasn't shy about proclaiming its virtues to the world. The "E" battery would ensure that there would soon be "a miniature dynamo in every home . . . an automobile for every family." Edison declared that "the time has nearly arrived when every man may not only be able to light his own house, but charge his own machinery, heat his rooms, cook his food by electricity without depending on anyone else for these services." In other words, mankind would be freed from the wires that delivered AC to the home. The future belonged to portable DC power, available to anyone with an Edison battery.

Edison's declaration of victory, however, proved to be premature. Almost as soon as the "E" batteries went into service, troubling reports came back about their performance in the field. The batteries were leaky, the electrical contacts often failed, and the units tended to lose power quickly, especially in cold weather. Edison was mortified. The batteries carried the Edison name, which meant more to him than anything. Against the advice of his business advisers, Edison immediately recalled the "E" batteries and shut down production, all at considerable expense.

The recall would have ruined most manufacturers, but Edison was able to draw on the resources of his other operations. Edison's 1903 film, *The Great Train Robbery*, filmed shortly after the electrocution of Topsy, turned out to be a huge hit, the first blockbuster movie. The profits from the film kept Edison's floundering battery business afloat, and the inventor set out more determined than ever to build a better battery.

"In phonographic work we can use our ears and our eyes, aided with powerful microscopes," Edison said. "In the battery our difficulties cannot be seen or heard, but must be observed by our mind's eye."

Edison wasn't hearing much of anything by this point. In 1905, he underwent an operation for mastoiditis, which robbed him of what little hearing he had. By now, he was entirely deaf in his left ear and severely impaired in his right one. Many photos of Edison later in life show the inventor cocking his "good" ear to a visitor, straining to make out what was being said.

Still, Edison pushed on. One of his employees recalls, "Sometimes, when Mr. Edison had been working long hours, he would want to have a short sleep. It was one of the funniest things I ever witnessed to see him crawl into an ordinary roll-top desk and curl up and take a nap. He would use several volumes of *Watts's Dictionary of Chemistry* for a pillow, and we fellows used to say that he absorbed the contents during his sleep, judging from the flow of new ideas he had on waking."

To stimulate his employees, Edison hung a large sign in his laboratory that showed how various compounds had fared in battery trials. After more than ten thousand additional experiments between 1905 and 1909, Edison came up with yet another battery design. He developed a process to make the positive electrode out of thin nickel flake by alternately electroplating layers of copper and nickel on a metal cylinder and then dissolving the copper away in a chemical bath. The flakes were arranged so that small charges of nickel hydrate and nickel flake were alternately layered into the pockets of the positive electrode, and then tamped down under tremendous pressure, about four tons per square inch. This ensured near-perfect contact and excellent electrical conductivity throughout the entire battery, while keeping the weight of the battery low. Ten years and more than a million dollars after he began, Edison's storage battery was ready to power the world.

Edison proclaimed the new "A" battery as being "almost a perfect instrument."

He launched splashy promotions, putting a battery-powered electric car on a grueling 1,000-mile endurance tour. Magazine and newspaper ads proclaimed, "THIS BATTERY WILL OUTWEAR YOUR CAR." A 1909 ad for a Detroit Electric car equipped with an Edison battery announced, "The success of the Detroit with the Edison battery has passed even the expectations of its inventor. Next season, an electric not thus equipped will be as out-of-date as a single-cylinder gas engine."

The Edison battery extended the range of electric vehicles to one hundred miles between charges; in one test a Detroit Electric car traveled more than two hundred miles on a single charge. The battery required little care—all a motorist had to do was fill the battery once a week with water and renew the electrochemical solution once a year. The "A" battery was lighter than equivalent lead-acid batteries, could be recharged in half the time, and lasted at least three times longer.

Edison saw the battery as not simply the salvation of the electric automobile but also the belated vindication of his beloved DC standard. For a while, it looked as though the Wizard had pulled off an unlikely comeback. Beginning in 1910, electric vehicles equipped with Edison batteries enjoyed brisk sales. Electric cars didn't require gasoline, which was expensive by the standards of the day, and were easier to start than autos with internal combustion engines, which had to be started with a hand crank. The top speed of the electric car was about twenty miles an hour, but the poor condition of most roads at the time made going much faster impractical anyway. Detroit Electric was selling close to two thousand electric cars a year, most of them powered by Edison batteries.

Flushed with success, Edison looked forward to the day when DC would power not only electric cars but also the engines of heavy industry. AC power plants would eventually become something akin to filling stations, used to recharge Edison's DC batteries. In the AC/DC war, the victor would become the vanquished, and Edison would be proven right after all. In a 1910 article for *Popular Electronics*, Edison wrote, "For years past I have been trying to perfect a storage battery, and have now rendered it entirely suitable to automobile and other work. Many people now charge their own batteries because of lack of facilities, but I believe central stations will find in this work very soon the largest part of their load. The New York Edison Company or the Chicago Edison Company should have as much current going out for storage batteries as for power motors, and it will be so some near day."

But it wasn't to be. The market for electric cars peaked just two years after Edison's storage battery came on the market and then went into a tailspin, a victim of changing technology and market conditions. Once again, Edison had backed the wrong horse. The invention of the automobile electric starter in 1912 eliminated the need for the hand crank for internal combustion engines, making gasoline-powered cars as easy to start as electrics. The discovery of Texas crude oil dramatically reduced the price of gasoline, making it affordable for the average working person. Improved roads made high-speed and long-distance travel more practical, putting electric cars—with their limited speed and range—at a distinct disadvantage. When Henry

Ford developed the first inexpensive mass-produced gasoline cars in the early 1920s, it was the final blow for the electric car. Electric vehicles all but disappeared by the mid-1930s. It would be half a century before anyone seriously took them up again.

Edison's storage battery turned out to be well suited for many applications, just not for the needs of powering a thousand-pound car sixty miles an hour. The Edison battery would continue to be used to power railway signals, industrial machinery, electric hand trucks, miner's lamps, and as power backup for AC. For the last twenty years of Edison's life, batteries would be his most reliable moneymaker. But the storage battery would never enjoy the sort of widespread use that Edison had envisioned. DC would have to take a back seat to AC—again.

George Westinghouse wasn't gloating over his victory. In 1907, Westinghouse became badly overextended during one of the periodic financial panics that swept the country around the turn of the century. His creditors forced him to resign from the Westinghouse Electric and Manufacturing Company and turn over control of the company to a consortium of bankers. Westinghouse was so distraught about losing his company that for years, whenever his train passed the Westinghouse Electric plant in Pittsburgh, he would turn his head away.

After losing his company, Westinghouse returned to his first love, invention. He developed a rotary steam engine for maritime use that quickly supplanted reciprocating engines in large ships, eventually becoming a worldwide standard. Westinghouse also came up with an improved compressed-air shock absorber for automobiles, a design similar to today's auto shock absorbers.

Westinghouse and Edison would never revisit their bitter war of the currents, nor would Edison make amends for the excesses committed in his name. But three years before Westinghouse died, he received an indirect tribute from Edison. In 1911, Westinghouse received the Edison Medal, an award that had been established by associates of Edison to honor groundbreaking achievements in the electrical arts. The Edison Medal citation praised Westinghouse "for meritorious achievement in connection with the development of the alternating current system for light and power." At the awards ceremony, it was noted: "It is perhaps somewhat ironic that he whom we are to honor tonight has disagreed violently over a long period of years with the man in whose honor this award was founded. But those of us who know Thomas Edison as a generous and just man know that he regards his defeat in one battle as a great victory in the march toward progress." However, there's little evidence that Edison had become magnanimous in defeat. He offered no congratulations to Westinghouse and declined to comment publicly about the award.

By this time, Westinghouse's health was in decline, and he retired from work in 1913. That winter, he caught a cold he couldn't shake, and his heart was so weakened that he was confined to a wheelchair. On the morning of March 12, 1914, George Westinghouse was found dead in his bed at age sixty-seven. Nearby were sketches of a new invention he was working on: an electric wheelchair. The chair was to have been powered by direct current.

"WESTINGHOUSE IS SUMMONED BY DEATH" announced the *San Francisco Chronicle*, adding a dull but fitting subhead: "Life was one of usefulness." Westinghouse was buried with full military honors in Arlington National Cemetery, and tributes flooded in from all over the world. One came from Nikola Tesla: "George Westinghouse was, in my opinion, the only man on this globe who could take my alternating-current system under the circumstances then existing and win the battle against prejudice and money power. He was one of the world's true noblemen, of whom America may well be proud and to whom humanity owes an immense debt of gratitude.

"I like to think of George Westinghouse as he appeared to me in 1888," Tesla recalled, speaking of the most ferocious year of the AC/DC battle. "The tremendous potential energy of the man had only in part taken kinetic form, but even to a superficial observer the latent force was manifest. He enjoyed the struggle and never lost confidence. When others would have given up in despair, he triumphed."

Even though Westinghouse had lost control of his company, his estate was valued at $50 million at his death. He was awarded 361 patents in his lifetime, including one that would be issued four years after his death, for an automatic train controller. Above all, Westinghouse's greatest and most lasting contribution was that he had utterly transformed the electricity business, overturning the locally distributed DC system with the long-distance AC system.

The Westinghouse name would live on; Westinghouse Electric and Manufacturing Company became an industrial powerhouse in the twentieth century. Westinghouse launched the first commercial radio station in 1920, KDKA, in Pittsburgh. Westinghouse generators powered the great hydroelectric projects, while its refrigerators, stoves, and washing machines quietly hummed away in millions of homes. Everyone knew the company's slogan: *You can be sure if it's Westinghouse*.

Westinghouse's accomplishments were so wide-ranging that many social critics believed that he would go down in history as one of the greatest inventors and businessmen of all time. A Westinghouse biographer wrote in 1922: "A thousand years from now, when scholars and philosophers try to measure the influence in the history of the human race in the era of manufactured power, and when they try to name the illustrious men of that era, they will write high in the shining list the name of George Westinghouse."

But the memory of George Westinghouse, far from enduring a thousand years, would barely last a generation. The man and his achievements would quickly fade from public view. Much of that was due to Westinghouse's own self-effacing nature. He gave only occasional interviews, wrote few private letters, kept no journals or notebooks, and left no significant store of papers behind. Not a single foot of movie film showing George Westinghouse has survived. In his fierce war with Edison, he had never personalized the battle. Westinghouse made no boasts, and he attracted little attention to himself. All he did was win.

Westinghouse's chief collaborator, Nikola Tesla, found it hard to capitalize on his AC triumph. Tesla's peculiar nature made him a solitary man, a loner in a field that was becoming so complex that it demanded collaboration. While Edison had a flair for inspiring men to work together for a common goal, Tesla had no such executive talents. Tesla found it difficult to work with others or to delegate responsibility. He was a man unto himself, unwilling or unable to reach out to others for inspiration or support. After his groundbreaking work on the polyphase AC motor, Tesla would continue to show flashes of brilliance. But as a businessman, he would be an utter failure.

"The inability to work with others, the inability to share his plans, was the greatest handicap from which Tesla suffered," wrote John O'Neill, Tesla's associate and first biographer. "It completely isolated him from the rest of the intellectual structure of his time and caused the world to lose a vast amount of creative thought which he was unable to translate into complete inventions. . . . Many scores of important inventions have undoubtedly been lost to the world because of Tesla's intellectual hermit characteristics."

After his work with alternating current, Tesla developed what came to be known as the Tesla coil, an induction coil that produced high-frequency, high-voltage currents. Tesla envisioned the coil as a wireless means of transmitting power, sending perhaps millions of volts around the globe. But the coil never saw commercial use and remains a curiosity.

Tesla was also an early pioneer in the emerging field of radio. In 1898, he patented a radio-controlled boat, which he demonstrated, to great acclaim, at an exhibition in Madison Square Garden. He followed up with a series of radio patents that would later pave the way for worldwide radio transmission. With surprising insight, Tesla predicted that the radio would one day be "a cheap and simple device, which might be carried in one's pocket. . . . And it will record the world's news or such special messages as may be intended for it."

But Guglielmo Marconi, a wealthy Italian nobleman and inventor, soon stole Tesla's thunder in the field of radio. In 1901, Marconi transmitted and received radio signals across the Atlantic Ocean for the first time. Three years later, the U.S. Patent Office awarded Marconi a patent for the invention of key radio components. Marconi's company became an immediate hit on Wall Street, and Edison eventually invested in the company and became a consulting engineer. Tesla was left with a handful of radio patents and little else to show for his efforts. He sued the Marconi Company for infringement in 1915, but lacked the resources to litigate the case. Eventually, the U.S. Supreme Court, in 1943, would rule that Tesla's initial radio patents had precedence over those of Marconi. By then it was too late. Marconi was forever "the father of radio."

Tesla's failure to capitalize on his inventions left him in almost constant financial turmoil. In 1917, Tesla joined George Westinghouse as the latest winner of the Edison Medal, "for meritorious achievement in invention and development of alternating current systems and apparatus." But soon, Tesla was so strapped for cash that he tried to sell the Edison Medal just to settle his unpaid bills. He was talked out of parting with the award, but wound up being evicted from his apartment at the Hotel St. Regis in New York anyway. For the last two decades of his life, Tesla lived alone in a series of New York hotels.

Tesla and Edison never crossed paths in their later years. Edison, by then, was an acclaimed international celebrity; a wealthy inventor and certified genius. Tesla was an eccentric whose accomplishments were far less tangible. In occasional newspaper interviews, Tesla made digs at Edison and his growing legend, often saying that Edison received more ideas from his associates than he contributed himself. Tesla was somewhat bitter about how things had turned out. Edison had help from hundreds of workers and had become a comfortable millionaire. Tesla had done practically everything himself, with little to show for it.

As Tesla's associate John O'Neill wrote, "Tesla was probably very unfair to Edison in this respect. The two men were entirely different and distinct types. Tesla was totally lacking in the university type of mind; that is, the mind which is adapted to cooperate with others in acquiring knowledge and conducting research. He could neither give nor receive, but was entirely adequate to his own requirements. Edison had more of the cooperative, or executive, type of mind. He was able to attract brilliant associates and to delegate to them major portions of his inventive research projects."

In later years, Tesla's behavior became increasingly erratic. On his seventy-eighth birthday, he announced in an interview that he had developed a "death ray" powerful enough to destroy 10,000 airplanes at a distance of 250 miles, and wipe out an army of a million men instantly. He also claimed to have designed a device that could generate gigantic tidal waves, which the U.S. Navy could use to sink enemy ships. He was also working on a "telautomaton," a self-propelled machine that could be controlled by impressions received through the eye.

Tesla took up residence at the Hotel New Yorker in Manhattan, the largest hotel in New York at the time, and developed a peculiar obsession for feeding pigeons. He was a familiar figure on the plazas of the New York Public Library and St. Patrick's Cathedral, clutching a sack full of bird feed. Tesla would shuffle out to the center of the plaza and let out a low whistle; pigeons would appear from all directions, carpeting the sidewalk and even perching on his shoulders. If Tesla was unable to make his daily rounds, he would summon a Western Union messenger boy, pay him his fee plus a dollar tip, and send him out to feed the birds.

Tesla left the windows of his hotel room open so birds could flutter in and feed. In 1937, Tesla was struck by a taxicab while crossing the street, breaking three ribs and severely injuring his back. He was bedridden for months, but as soon as he was on his feet, he was back feeding the pigeons.

"This devotion to his pigeon-feeding task seemed to everyone who knew him like nothing more than the hobby of an eccentric scientist," wrote John O'Neill. "But if they could have looked into Tesla's heart, or read his mind, they would have discovered that they were witnessing the world's most fantastic, yet tender and pathetic love affair."

On January 7, 1943, Tesla was found dead in his bed in Room 3327 of the Hotel New Yorker. Nearly two thousand people attended his funeral service at St. Patrick's Cathedral. His pallbearers included executives from General Electric and Westinghouse. The factions that had once fought so bitterly over AC and DC were now on opposite sides of a casket.

Two days after Tesla's death, agents from the U.S. Office of Alien Property broke into his hotel room and seized his possessions, including a cache of technical papers. Tesla's frequent talk of working on a "death ray" had not gone unnoticed by the U.S. military. With World War II raging, the government was anxious to secure Tesla's papers to prevent them falling into enemy hands. A government scientist spent several days poring over Tesla's papers and concluded that while they contained a great deal of speculation about the wireless transmission of power, they "did not include new, sound, workable principles or methods for realizing such results."

An FBI report filed shortly after Tesla's death was more blunt: "Concerning Tesla, hotel managers report he was very eccentric if not mentally deranged during the past ten years and it is doubtful if he has created anything of value during that time." Still, the tantalizing possibility that Tesla had come up with a beam weapon would keep the U.S. government interested in Tesla's work for decades.

Shortly after World War II, a U.S. Air Force operation codenamed "Project Nick" (for Nikola) studied the feasibility of beam weapons, using Tesla's papers as inspiration. As late as the 1980s, the Air Force showed an active interest in Tesla's work, hoping that Tesla's beam weapons research would aid the Strategic Defense Initiative, the space-based antimissile system dubbed "Star Wars." In February 1981, an Air Force lieutenant colonel working on SDI wrote a memo to FBI director William Webster that said in part: "We believe that certain of Tesla's papers may contain basic principles which would be of considerable value to certain ongoing research within the Department of Defense. It would be very helpful to have access to his papers."

But Tesla's death beam was no more real than Star Wars turned out to be. Tesla's reach had always exceeded his grasp, but in his later years, more than a few of his proposed inventions skirted the blurry boundary between genius and madness. Tesla's best invention turned out to be his first: the induction motor and the AC polyphase system. If the rest of Tesla's career didn't quite measure up, it didn't matter: once, he had caught lightning in a bottle.

Of all the combatants in the AC/DC struggle, it would be the loser who would fare best. For Thomas Edison, the defeat of his cherished DC standard was a bitter blow, but hardly fatal. The inventor was involved in far too many other projects for any one setback to derail him financially. However, Edison never again took up electricity as a serious experimental subject. He had tasted defeat in the field once, and that was quite enough. "People will forget my name ever was connected to anything electrical," Edison said late in his life. It was less a prediction than a wish.

Edison continued to refine his storage battery even after it was clear that batteries would never become a primary source of power for automobiles or industry. Some of Edison's associates drifted away to form new companies. Soon, his laboratory in Orange was gone, too. On the evening of December 9, 1914, a fire broke out in a section of his lab where he stored flammable chemicals and film stock for his movie company. Gigantic flames leaped from the building, igniting adjacent structures, and turning the entire laboratory grounds into a raging chemical inferno. All of Edison's work was literally going up in smoke, but the inventor couldn't help but be impressed by the tremendous power of the blaze. Watching the lab burn, Edison told his son Charles to fetch his wife immediately. "Get her over here, and her friends too," Edison said. "They'll never see a fire like this again."

For many years, Edison would have nothing to do with General Electric, the company that was sold from under him. But in his seventies, mellowing somewhat with age, Edison relented, journeying to Schenectady in the fall of 1921 to tour GE's massive plant and corporate headquarters. He had been away from the company nearly a quarter century, and he scarcely recognized the place. GE was now one of the country's leading suppliers of equipment for producing alternating current dynamos, transformers, and transmission lines.

The company had expanded into building airplane engine boosters for the fledgling aviation industry, as well as refrigerators and stoves for the home and X-ray machines for hospitals. Edison seemed to enjoy his tour of GE, but he would later complain about the rise of the corporation laboratory, with its cautious bureaucracy. The best inventions, Edison believed, sprang from laboratories where one man's vision ruled over a team of talented assistants. To Edison, General Electric was all soldiers and no general. But Edison's ways were passing. The number of patents awarded to corporations exceeded those granted to individuals for the first time in 1931.

In his final years, Edison returned to an issue that had preoccupied him as a youth—education. He had always railed against the formal education he never received, insisting that people were better served by developing their own powers of critical thought. Now he believed more strongly than ever that higher education was a waste of time and money. Colleges needed to be teaching people how to think, not cramming their minds with useless facts. The only educational philosophy he favored was the Montessori Method, which held that children should be free to learn without restriction or criticism.

"I wouldn't give a penny for the ordinary college graduate, except those from institutes of technology," Edison declared. "They aren't filled up with Latin, philosophy, and all that ninny stuff. America needs practical skilled engineers, business-managers and industrial men."

Edison even devised an IQ test of his own, which he administered to prospective employees. He dubbed his test the "Ignoramometer." The 150 questions were idiosyncratic, like the man himself: What voltage is used on streetcars? (Six hundred volts at the time.) Which countries supply the most mahogany? (Brazil and Bolivia.) Where is Magdalena Bay? (Baja California.) How many cubic yards of concrete are in a wall 12 feet by 20 feet by 2 feet? (It works out to 17.78 cubic yards.) Who was the Roman emperor when Jesus Christ was born? (Augustus.) Many of the questions were passed along in newspaper articles, and the Ignoramometer stirred a lively debate over whether Edison was right about measuring intelligence—or just getting cranky in his old age.

"Of course, I don't care directly whether a man knows the capital of Nevada, or the source of mahogany, or the location of Timbuktu," Edison said. "But if he ever knew any of these things and doesn't know them now, I do very much care about that in connection with giving him a job. For the assumption is that if he has forgotten these things he will forget something else that has direct bearing on his job."

In 1927, at the age of eighty, Edison announced his official retirement from experimenting. He still gave annual interviews to newspapers on his birthday, weighing in on everything from Soviet Russia ("Everything there is like a machine and nobody likes it") and his favorite invention (the phonograph, because "it has brought so much joy into millions of homes") to the future of war ("Future wars are going to be waged almost exclusively with airplanes, submarines, and gas. Battleships will not count for much.")

In all his public pronouncements, however, the proud Edison would never discuss his greatest setback, losing the war of the currents. He would never admit that he had waged the fight by less than honorable means, or that his actions were fueled by emotion and spite rather than the reason he always championed. Reporters knew not to even bring up the subject. It was enough that practically every light bulb, industrial motor, and appliance in America was powered by alternating current. The proof of Edison's mistake was humming all around him. Even a deaf man could hear it.

Only once did Edison admit his error about AC, and that was privately. In 1908, Edison happened to meet the son of William Stanley, the one-time chief engineer for George Westinghouse who designed the first-ever AC transmission system in the United States, in Great Barrington, Vermont. Edison motioned for the young Stanley to come closer. "Oh, by the way," Edison told the youth in a low voice. "Tell your father I was wrong."

Edison studiously avoided anything that reminded him of the war of the currents, particularly his chief hatchet man at the time,

Harold Brown. Edison and Brown parted ways after the William Kemmler execution and never worked together again despite Brown's best efforts to ride Edison's coattails. In 1902, Brown was promoting himself as the exclusive agent of "The Edison-Brown Plastic Rail Bond," a conductive bonding agent for iron rails he supposedly co-invented with Edison. When Edison learned of Brown's claim, he quickly dispatched his lawyers. Brown was forced to drop the use of Edison's name in his company's letterhead and advertisements. Three years later, Brown was at it again, this time attempting to register a trademark for a conducting agent he wanted to call "Edison Solid Alloy." Edison's lawyers once more intervened by contesting the trademark with the U.S. Patent and Trademark Office, Edison himself writing that Brown "had not been entirely ingenuous in his relations with me." Once again, Brown was compelled to stop using Edison's name.

Still, Brown never stopped reminding anyone who would listen that he had once "worked" for Edison, a far cry from the days when he hotly denied ever being in Edison's pay. In 1918, Brown was one of the founding members of the Edison Pioneers, a group of former Edison employees. Brown professed to have worked with Edison as far back as 1876, which, had it been true, would have made Brown one of the inventor's first employees. By the 1940s, there were only five surviving Edison Pioneers. One of them was Harold Brown, still basking in the reflected glow of the man who brought light to the world.

In his twilight years, Edison was celebrated as a national treasure, a symbol of American ingenuity. He was awarded a Congressional Medal of Honor in 1928 for his lifetime of invention. The following year, Edison took ill, collapsing during a jubilee held to mark the fiftieth anniversary of his incandescent lamp. He was confined to an easy chair at home, but still kept up with the workings of his laboratory. In January 1931, Edison was awarded U.S. patent number 1,908,830, "A holder for an article to be electroplated." It was the 1,093rd patent of his career, and his last.

That summer, Edison's health took a turn for the worse, and for several months, he hung between life and death. In the early morning hours of October 18, 1931, Edison was found dead in his bed at his estate in West Orange, New Jersey. The light that had burned so brilliantly was at last extinguished.

Edison's body lay in state for two days in the library of his laboratory, surrounded by mementos of his many triumphs. "WORLD MADE OVER BY EDISON MAGIC" said the *New York Times*, a sensational headline that happened to be absolutely true.

A group of Edison admirers pushed for a unique tribute: on the day of Edison's funeral, all electric current in the country would be shut off for two minutes. But the proposal drew immediate criticism from businesses and factory owners who argued that cutting the power would cost tens of millions of dollars in lost production. Edison was born into a world without electrical power; now, thanks to his invention, the world couldn't do without electricity for two minutes. As a compromise, on the day of Edison's funeral, lights all across the country were voluntarily dimmed at 10 P.M. At 10:02, the lights came up again, every one of them powered by the alternating current Edison had fought so hard to discredit.

12

DC'S REVENGE

With Edison gone, DC had lost its greatest champion. By then, it hardly mattered. Practically everything was running on alternating current by the 1930s—generators, motors, and electrical devices—and the investment in the AC standard was so colossal that there was no turning back. Its dominance seemed assured. Many of the one-time opponents of AC—chief among them the company Edison founded, General Electric—became its most enthusiastic champions. AC wasn't simply portrayed as a safe and efficient way of distributing electricity; it was nothing less than a godsend.

"ELECTRICAL LIVING . . . THE PROMISE OF THE FUTURE" proclaimed a 1944 GE advertisement. The overheated copy read: "Electricity has woven itself so inseparably into our lives that its miracles are taken for granted. Its sleepless power leaps to our fingertips to perform tasks, which only yesterday etched youthful faces and lovely hands with the indelible lines of toil and fatigue. Its tireless energy takes the place of yesterday's human effort. Today, women are awakening to electricity as preserver of youth, giver of freedom. . . . Womankind gratefully turns task after task over to electricity, her obedient and faithful servant, and quickly adapts herself to a richer, happier life—the NEW ELECTRIC WAY!"

General Electric became one of the twentieth century's most profitable corporations by successfully selling the public on the promise of the all-electric Utopia, a life made easier by the quiet servant humming in the wires. GE and Westinghouse Electric had large investments in both power generation and electrical appliances, so by promoting the use of more electric gadgets, the companies also increased demand for electricity, further stoking their profits. By 1940, Westinghouse had annual sales of more than $400 million; GE had more than $1 billion.

The demand for electricity spiked sharply during World War II as the country shifted into war production. When the shooting stopped, the demand for electricity kept growing. The Baby Boom and the move to the suburbs, in many respects, were powered by alternating current. The amount of electricity generated by U.S. utilities in the 1950s and 1960s increased by nearly 10 percent annually, and to keep up, the nation's power distribution network became increasingly more complex and interdependent.

AC's greatest triumph was how quickly it rendered itself invisible, quietly receding into the walls, out of sight and out of mind. In the second half of the twentieth century, electricity became something everyone took for granted, noticed only when something went wrong. The occasional power failure would force people to realize how dependent on electric power modern life had become. But in short order, the power would be restored, and electricity would slip back into the walls, invisible and unnoticed.

The gigantic distribution network that electrified the continent—the North American power grid—grew to become the largest machine ever built by humans. The configuration of the grid would be shaped by the nature of AC itself, the relative ease with which it can be transmitted long distances cheaply.

Today, the North American power grid consists of four massive subsystems, each providing alternating current to a different section of the country. The Eastern Interconnect supplies power primarily to users east of the Rocky Mountains, the Western Interconnect handles customers west of the Rockies and portions of northern Mexico, the Quebec Interconnect covers that Canadian province, and the Texas Interconnect serves Texas and bordering states. As a result, electricity often flows great distances before it is consumed. A turbine spinning in Ontario may power a light bulb in New York; a television in Los Angeles draws on electricity generated in Montana. A single localized shipment of electric power spreads spider-like through a large section of the grid, altering flows on many other lines. Sending power through the grid from Wisconsin to Florida, for example, can alter the flow in dozens of adjacent states. The way electricity was first explained to Edison wasn't far off the mark: electricity was like a long dog with its tail in Scotland and its head in London—when you pulled its tail in Edinburgh, it barked in London.

AC triumphed in the twentieth century because it allowed electrical distribution to be centralized into a vast interdependent network. But in the twenty-first century, AC's strength is shaping up to be its greatest weakness—the vulnerabilities of a large centralized network for generating and distributing electricity are already beginning to show. Despite the safeguards built into the North American grid, local power outages periodically have a cascading effect, plunging an entire region into darkness and chaos. On November 9, 1965, the all-electric Utopia had its first major meltdown in what became known as the Great Northeast Blackout. It was the largest power outage in history, affecting nearly 30 million people in New York, New England, and Pennsylvania. Striking at the evening rush hour, the power failure trapped 800,000 riders on New York City's subways, halted railroads, snarled traffic, and left planes circling over airports. The cause of the blackout was the failure of a single transmission line relay.

As a result of the Great Northeast Blackout, additional safeguards were built into the grid in an effort to prevent outages from spreading. But large blackouts would prove to be almost impossible to eliminate entirely. In July 1977, another Northeast blackout left about 9 million people in New York City without power for more than a day, and there was looting in the darkened streets. There were widespread outages in the Western Interconnect in 1994 and 1996, and in the Eastern Interconnect in 1999. In August 2003, another major power outage rippled across the Northeastern United States and Canada, raising national security alarms and prompting calls to re-engineer the grid. In the age of terrorism, a large centralized electrical network makes for an inviting target. Repairing damage to a centralized electrical network is complicated and time-consuming. Three years after the U.S. invasion of Iraq, electrical production in that country was still below its prewar peak.

In many instances, the grid has become too centralized; what's needed in the future are smaller decentralized electrical systems that are less vulnerable to widespread breakdown. One of the proposed solutions to fix the grid is an idea that would warm the heart of Thomas Edison: bring back DC. In fact, DC is already quietly seeping back into the AC grid. Direct current is being used to get around one of the major problems in transmitting AC power from one section of the grid to another: synchronizing the alternating peaks and valleys of current. All AC produced in the United States has a frequency of 60 cycles per second, and those alternating waves have to be in step with one another when electricity is sent from one sector of the grid to another. As a result, DC links are becoming an increasingly popular way to connect out-of-step sections of the AC grid, eliminating the need to synchronize frequencies.

A growing number of high-voltage DC links now connect sections of the grid, including an 850-mile DC line between the Pacific Northwest and Los Angeles. What makes such long-distance DC lines possible is a device that might have turned the tide for Edison had it been around during his lifetime: the high-voltage valve. Essentially, the valve performs the same function for a DC system that a transformer does for an AC network, allowing the voltage to be stepped up for long-distance transmission and stepped down for local use. Such valves were first developed for commercial use in the 1950s, and have since been improved considerably by fashioning them out of silicon.

High-voltage DC (known in the electricity trade as HVDC) is now used extensively in Europe to connect different countries' AC power systems. HVDC is also becoming the preferred method for sending electricity underwater by cable. Transmitting alternating current underwater builds up high capacitance, or stored electric charge, which has to be overcome with additional current; direct current is virtually unaffected by being transmitted underwater.

Scores of long underwater power lines now transmit direct current, among them a 155-mile cable that runs under the Baltic Sea connecting Sweden and Germany and a 67-mile DC line from northern New Jersey to central Long Island, New York. Wind farms are also turning to HVDC systems to collect power from a series of unsynchronized generators and transmit it by cable. A California utility is considering a 650-mile underwater DC transmission line that would relay wind and hydroelectric power from the Pacific Northwest, where power is relatively plentiful, to the San Francisco Bay Area, where energy sources are scarce. The link would be the world's longest undersea high-voltage DC line. If Edison had the same technology available at his Pearl Street power plant, he could have transmitted DC power to customers as far away as Cincinnati.

High-voltage DC may even turn out to have health advantages over AC. Some epidemiological studies have reported a link between exposure to the low-frequency electromagnetic fields that surround AC power lines and increased rates of leukemia and other cancers. The health risks remain largely unproven and are the subject of considerable dispute, but an increasing number of communities are fighting to keep new high-voltage AC lines from running through their neighborhoods. (Imagine what Harold Brown might have done with an issue like this.)

It's taken the better part of a century for direct current to creep back as a supplement to alternating current. The next hundred years may see DC taking on AC head-to-head once again as the demand increases for portable power. Every portable electronic device on the planet—laptops, cell phones, PDAs, MP3 players—already runs on direct current. The future of computing lies in making digital devices truly portable, so that users can communicate on any device, anytime, from anywhere in the world. To build the "always connected" world, devices will have to be untethered from wires, including the wall outlet, and powered by long-lasting rechargeable batteries or fuel cells. In short, a move from AC to DC. The Industrial Age was powered almost exclusively by AC, but the Computer Age may well turn out to be DC's revenge.

If Edison were alive today, he'd no doubt be in the thick of the effort to come up with a powerful and portable "box of electricity" to power electronic devices and even automobiles for days or weeks on a single charge. The fact is, batteries have improved only marginally since Edison's day. Although modern batteries are more durable and much less prone to leak, their performance hasn't kept up with advances in electronics. The average laptop battery, for example, holds a charge for only two to five hours, depending on what functions the computer is asked to perform. Even if battery performance were to double—which isn't likely to happen soon—a laptop battery still couldn't hold a charge for an entire workday.

Battery makers continue to tweak the chemical makeup of batteries to improve performance, in much the same way that Edison tirelessly experimented with hundreds of chemical permutations. But such efforts are likely to improve the longevity of batteries only marginally. The world awaits a dramatic breakthrough in technology to power a society increasingly dependent on portable electricity.

Battery life is now considered a critical bottleneck in the advancement of computers and consumer electronics. A recent survey of consumers in fifteen countries revealed that the single most desired feature in a future mobile device was a longer-lived battery. The survey also showed that poor battery life on mobile devices was one of the main reasons people did not use their portable gadgets more often.

The most promising new technology that could dramatically improve portable DC power is the fuel cell. Fuel cells are essentially batteries with a refillable energy source, usually hydrogen, the simplest of all elements. In a fuel cell, electrons are stripped from the hydrogen, resulting in a flow of electrical current, and the remaining hydrogen ions combine with oxygen to form water, the only byproduct of the hydrogen fuel cell. Unlike secondary batteries, which have to be regularly recharged with electricity, fuel cells can produce power indefinitely as long as they are supplied with hydrogen and oxygen.

Fuel cell development still faces significant technical hurdles. Hydrogen, which doesn't exist freely in nature, costs $5 a gallon to extract and process, and the energy used usually produces greenhouse gases, making the fuel cell far from emission-free. Before fuel cell cars can be truly competitive with gasoline-powered autos, a massive infrastructure of hydrogen plants and fueling stations will have to be built.

Still, it's a good bet that one of Edison's most ambitious projects, a box of direct current powerful enough to run a car, will come into its own in the decades ahead, eventually bringing to an end the reign of the internal combustion engine. In the long run, Edison may not have been so much wrong about DC as 150 years ahead of his time.

So it is with standards wars; all victories are provisional, all defeats subject to revision. Advances in technology, changes in the marketplace, in the way people live, and most important, in what they value, can overturn even the most entrenched technical standard. This has turned out to be especially true of electricity, built, as it is, on a foundation of dualities: negative and positive, AC and DC. What was bad becomes good, what was good becomes obsolete, the pair of opposites eternally alternating.

Epilogue

STANDARDS WARS

Past, Present, and Future

In a standards war, the dead often outnumber the living.

The vanquished include not only the companies that sponsored a failed standard but also the customers who bought the now-obsolete products from them. And it doesn't end there; even consumers who pick the winning standard wind up getting shortchanged. Prices remain artificially inflated well after a standards war is over, while the winning companies enjoy a temporary stranglehold on the market. In the end, a standards war only truly benefits a handful of big-money concerns, while consumers wind up footing the bill.

Standards wars are fought over technical distinctions, but, as in the AC/DC dispute, the conflict almost always goes deeper. They are skirmishes in a larger war between rivals in an emerging industry, a fight not simply for control of the market but for control of future markets. Fortunes and reputations are on the line, so it's not surprising that such disputes often become a clash of egos as much as standards. Leaders of the warring parties come to view their standard as an extension of themselves; they would no sooner abandon it than they would cut off an arm. As a result, many losers in standards wars go down to ignominious defeat, desperately clinging to their original idea long after it's clear to everyone else that they've lost.

The technical distinctions in a standards battle are often minor compared to the larger war for market dominance in an industry. That's certainly the case in the latest standards dispute to visit the

Digital Age, between two high-definition DVD standards known as Blu-ray and HD DVD. Technically, the two sides aren't very far apart. The *Blu* in Blu-ray refers to the blue-violet laser that's used to read and write data, a short-wavelength beam that allows the disk to store substantially more information than a standard DVD. But the rival HD DVD format also employs a blue laser, at the same wavelength as Blu-ray. Apparently, there's room for only one true-blu standard. (It's spelled *Blu,* incidentally, because "Blue ray" was deemed too common a phrase to be trademarked.)

The main technical differences between the two high-definition DVD standards have to do with the storage capacity of the disks and the cost of producing them. A single-layer Blu-ray disk holds about four hours of high-definition video with audio; HD DVD holds considerably less, about two and a half hours of video and audio. More expensive multilayer versions of both disks could potentially hold even more data; a four-layer Blu-ray disk has been proposed that could hold more than fifteen hours of high-definition video.

While HD DVDs can't store as much data, the disks themselves will likely be cheaper to manufacture than Blu-ray, as will HD DVD players, at least initially. The HD DVD camp is betting that consumers will be willing to trade some storage capacity for lower cost. The Blu-ray companies believe that future video applications will demand disks with very large storage capacities, and that consumers will be willing to bear the added expense.

The differences between the DVD standards are insignificant compared to the larger war for control of the home electronics market being waged among Sony, Blu-ray's primary backer, and Toshiba, Microsoft, and Intel, the triumvirate pushing HD DVD. Sony and Microsoft already square off head-to-head in the lucrative video game console market; the DVD standards dispute is a kind of proxy war for the larger battle.

Already, the DVD battle has featured many of the elements found in most standards wars of the past: the division into camps based on existing corporate rivalries rather than on technical merit, the increasingly shrill claims made by opposing sides, and the appeals to fear. (Don't get left behind!) Both camps are lobbying movie studios to adopt their standard, while dropping dark hints to customers about becoming stuck with the "wrong" format.

It's a routine that at least one of the main players—Sony—knows well. The Japanese company was one of the main combatants in a strikingly similar standards war a generation ago, which pitted Sony's Betamax videotape standard against rival VHS for control of the then-emerging home video market. In that battle, the technical distinctions between the two standards were also rather slight. The two formats were slightly different sizes—VHS cartridges were about an inch and a half longer than a Betamax—and they ran at different speeds—Betamax ran slightly faster. VHS's larger tape shell and slower running time meant that it could hold twice as much tape—two hours worth of programming, compared to Beta's one-hour limit. Betamax traded playing time for picture quality—the video image of a Betamax tape was somewhat sharper than VHS, and the way the tape was wrapped around the heads—vaguely in the shape of the Greek letter Beta—kept the tape threaded more securely around the video heads, allowing for faster and more precise tape cueing.

When Betamax debuted in November 1975, Sony's proud patriarch, Akio Morita, boldly declared it to be a standard for the ages, one that would launch a revolution in watching video at home. The revolution wouldn't come cheap—one of Sony's first U.S. offerings was a 19" color TV/Betamax VCR console that retailed for $2,295, while a stand-alone Betamax player-recorder went for $1,260.

In 1976, a consortium of Sony competitors led by JVC launched a rival videotape standard, VHS. Sony's Morita poured scorn on the VHS format, dismissing it as an inferior knock-off of his company's more technically elegant system. Even though the VHS cartridge did have the advantage of being able to hold twice as much programming, Sony wasn't worried about the difference. Since most

American TV programs were an hour or thirty minutes long, the company reasoned that consumers wouldn't care about storage capacity as long as they could record their favorite show on a single Betamax cartridge. Sony bet that consumers wouldn't want to sacrifice better picture quality for a longer-running tape—and it bet wrong. Betamax would indeed launch a revolution in home video, but the standard wasn't around long enough to reap the benefits.

Sony's mistake in judgment became clear almost as soon as the standards war began. Consumers soon began taping movies and sporting events on their home machines, programs that were far too long to fit on a single Betamax tape. Betamax's ace in the hole—its sharper picture quality—was only discernable on expensive TV sets; the average viewer couldn't see much difference in picture quality between the two standards. In effect, Sony was asking consumers to sacrifice tape capacity—something they quickly came to value as a critical feature—for improved picture quality that most of them couldn't see.

The VHS forces, meanwhile, lined up a small army of manufacturing partners and began churning out millions of VHS players, driving down the format's price. Consumers flocked to the cheaper, longer-playing VHS format, leaving Sony in the dust. The standards war was virtually over two years after it began; by 1978, VHS had a 70 percent share of the market, a lead it never surrendered. By the early 1980s, members of the original Betamax group, including Toshiba, Sanyo, and NEC, began selling VHS.

But like Edison and his cherished DC standard, Sony refused to admit defeat, even when confronted with the dire sales figures. By 1984, forty companies were making players utilizing the VHS format, compared to only twelve manufacturing Betamax decks, and Betamax's share of the consumer market had slipped to less than 20 percent. Betamax tapes became increasingly hard to find at retail outlets, further driving customers to the more popular VHS standard. In a last-ditch effort to salvage Betamax, Sony ran a series of newspaper ads with headlines posing provocative questions such as: "Is Betamax Dead?" "Is Buying a Betamax a Disadvantage?" and "What's Going to Happen to Betamax?" A final advertisement set the record straight: "Betamax: Getting More and More Exciting All the Time!"

But the only excitement left for Betamax was the spectacle of watching a multimillion-dollar standard going down in flames. In 1988, Sony finally conceded defeat and began producing VHS recorders, much as General Electric was finally forced to adopt alternating current. A Sony deputy president admitted the painful truth to employees, "Speaking frankly, we didn't want to manufacture VHS. However, you don't conduct business according to your feelings."

Sony made several key mistakes along the way with Betamax. The company was slow to license manufacturers to produce Betamax machines, handing VHS an early advantage on store shelves. Sony handled most of the research and development of the Betamax standard by itself, while the VHS standard was continually improved by dozens of competing manufacturers. And alone among the combatants, Sony let ego get in the way of further developing the standard, blindly defending its technology to the end. The single biggest problem with Betamax, the one-hour running time of the tape, was something few gave much thought to when the standard was launched.

It's surely no coincidence that in the latest DVD standards war, Sony is backing the longer-playing standard. Having been burned once by choosing the smaller, more technically elegant standard, Sony is putting its weight behind the "bigger-is-better" DVD standard. But with a standards battle, there's always the danger of fighting the last war rather than the current one. While it's likely that consumers will prefer longer-playing DVDs if given the choice, any number of other factors could trump that preference—lower price, better reliability, stronger industry support.

Losers in a standards war often cling to the notion that their standard is intrinsically superior, when in fact such claims are relative, subject to changing market conditions and shifting consumer whims. The better standard is simply the one adopted by the most people.

The DVD standards war will, if anything, discourage consumers from buying into either standard. Faced with two sets of high-definition video machines in the stores, many will prefer to wait out the battle in order to avoid being saddled with an out-of-date unit. The danger for the competing DVD companies is that in the rapidly evolving digital era, a new technology may come along and supplant both Blu-ray and HD DVD, resulting in a standards war with no winners at all. In the Museum of Rejected Standards, there's always room on the shelves for another small monument to human folly.

Further Readings in Electricity

By far the most revealing look at Benjamin Franklin's experiments with electricity was written by the man himself. *The Electrical Writings of Benjamin Franklin* compiles Franklin's personal letters and published works on electricity and is available free online through the Wright Center for Science Education at Tufts University (www.tufts.edu). The files are in the public domain, and can be searched, copied, and printed.

Thomas Edison was something of a pack rat, and his cache of personal papers is immense—5.5 million pages of documents, including correspondence, financial records, legal documents, manufacturing data, and newspaper and magazine clippings. Selections from the papers have been published in five hardback volumes, *The Papers of Thomas A. Edison*. Genius doesn't come cheap—the books retail for around $90 apiece—but much of the collection is available free online through Rutgers University (http://edison.rutgers.edu/). The bad news is the massive online archive is not a true electronic database and thus can only be searched in a very general way. Edison himself would love the find-the-needle-in-the-haystack aspect of his online papers, but for mere mortals, the digital collection can be frustrating to navigate.

For a more manageable introduction to the life and work of Edison, Matthew Josephson's *Edison: A Biography* (Wiley, 1959) is the classic standard biography, and still holds up. It's a bit uncritical in parts, but it does a nice job of capturing both Edison the man and his inventions.

Francis Jehl's *Menlo Park Reminiscences* (Edison Institute, 1938) is an endearingly fusty account by one of Edison's laboratory assistants at Menlo Park. It's not always reliable on chronology, but it offers a rare view over Edison's shoulder as he works in his lab. The *Diary and Sundry Observations of Thomas Alva Edison* (Abbey, 1968) is a collection of articles Edison wrote for popular magazines and newspapers of the day, along with a brief diary extract from 1885. Some of the pieces suggest the help of a ghostwriter, but the cranky opinions are undeniably Edison's own. Too bad this book is so short.

Edison's official biography, Frank Lewis Dyer and Thomas C. Martin's *Edison, His Life and Inventions*, published in 1910 with Edison's cooperation, is available free online through Project Gutenberg (www.gutenberg.org). The book is often more interesting for what it doesn't say, and for the way it reveals Edison hard at work fashioning his own legend.

For more on Edison's film work, nothing beats *Edison—The Invention of the Movies (1891–1918)*, a four-DVD set put out by the Film Center of the National Museum of Modern Art, Kino Video, and the Library of Congress. It's an astonishing compilation of 140 Edison films, from the first shorts filmed in Edison's "Black Maria" studio up to his company's last feature-length film in 1918, accompanied by more than two hours of commentary by film archivists and scholars. *Electrocuting an Elephant*, the 1903 Edison film depicting the killing of Topsy the Coney Island elephant, is included on disk one.

Nikola Tesla left behind a more modest paper trail than Edison, but an interesting one nonetheless. Tesla's autobiography, *My Inventions: The Autobiography of Nikola Tesla* (Hart Brothers, 1982), which was originally published in 1919 as a series of magazine articles, nicely reveals Tesla's scientific rigor and his mystic dreaminess. *Nikola Tesla: Colorado Springs Notes, 1899–1900* is Tesla's work diary from a year of experimenting, mostly on the wireless transmission of electricity. It's a bit technical in parts, but offers interesting insights into the way Tesla's mind worked. John J. O'Neill's *Prodigal Genius: The Life of Nikola Tesla* (Angriff Press, 1944), written shortly after Tesla's death by a science writer who knew Tesla, is very good at capturing Tesla in his peculiar and sometimes sad later years. Marc J. Seifer's *Wizard: The Life and Times of Nikola Tesla* (Citadel Press, 1996) is the best of the recent Tesla biographies, with new information about the FBI's continued interest in Tesla, even long after his death.

George Westinghouse, poor soul, had a pair of biographies written about him shortly after his death, *A Life of George Westinghouse* by Henry G. Prout (Scribner, 1922), and *George Westinghouse, His Life and Achievements* by Francis E. Leupp (Little, Brown, 1918), and hasn't been heard from much since. Westinghouse made history but left little of it behind.

George Westinghouse still roams the halls at the George Westinghouse Museum in Wilmerding, Pennsylvania, outside Pittsburgh (phone 412-825-3004; www.georgewestinghouse.com). The museum includes a full-size replica of a Westinghouse Time Capsule, a recording of the world's first commercial radio broadcast, and an Appliance Room full of early Westinghouse refrigerators, sewing machines, washers, and dryers. Ed Reis, the executive director of the museum, does a forty-five-minute program impersonating George Westinghouse for local groups (phone 412-655-2447, or e-mail ejreis@comcast.net).

The newly renovated Edison National Historic Site in West Orange, New Jersey (phone 973-736-0551; www.nps.gov/edis/) recreates the lab where Edison worked the last four decades of his life. Nearby, in the town of Edison, New Jersey (formerly Menlo Park), is the Menlo Park Museum (phone 732-549-3299), which contains an interesting collection of Edisonia, including vintage phonographs and wax recordings.

The Memorial to Topsy the Elephant is located in the Coney Island Museum, 1208 Surf Avenue, Brooklyn, New York (phone 718-372-5159).

The Author

TOM MCNICHOL is a contributing editor for *Wired* magazine. His articles have appeared in the *New York Times*, *Salon*, the *Washington Post*, and the *Guardian*. His radio commentaries and satires have aired on NPR's *All Things Considered*, *Morning Edition*, and *Marketplace*. He's the author of *Barking at Prozac* (Crown Publishing, 1997), and his work appears in the anthology *Afterwords: Stories and Reports from 9/11 and Beyond* (Washington Square Press, 2002). He and his wife, Tonia, live near San Francisco.

Index

F